# HARNESSED

How Language and Music Mimicked Nature and Transformed Ape to Man

## MARK CHANGIZI

D1502851

BENBELLA BOOKS, INC.
DALLAS, TEXAS

# Contents

# The Reading Instinct

At the beginning of his book *The Language Instinct*, Steven Pinker demonstrates the amazing power of language with an example. He writes:

The [language] ability comes so naturally that we are apt to forget what a miracle it is. So let me remind you with some simple demonstrations. Asking you only to surrender your imagination to my words for a few moments, I can cause you to think some very specific thoughts:

When a male octopus spots a female, his normally grayish body suddenly becomes striped . . .

Cherries jubilee on a white suit? Wine on an altar cloth? Apply club soda immediately . . .

When Dixie opens the door to Tad, she is stunned, because she thought he was dead . . .

With just a handful of words, our brains are pulled hither and thither to far-off corners of a vast mental universe, and new content is installed. For me, the Dixie-and-dead-Tad story from *All My Children* is old news, but a few of you may not have known Tad is alive. And now you know, from just a few words in the right order.

That kind of brainpower doesn't happen by accident, Pinker argues. The deeply malleable, blank-slate brains the social sciences have long supposed we possess could never learn and do language as we can. Language is astoundingly complicated—to this day, we cannot build effective speech-recognition machines—and yet we are uncannily good at it: children learn language too quickly and easily, we all comprehend it too automatically and effortlessly, and it pervades our life too completely to be something we simply learn with general-purpose brains. And our brains, indeed, have long appeared to have specialized regions for language. That we have an instinct for language is also suggested by its universality: language is found everywhere, and languages tend to share many common features.

And although Pinker may not extend these arguments to music—he famously called music "auditory cheesecake"—other researchers would. Steven Mithen, in *The Singing Neanderthals*, pointed out that music is complex, and yet we're creepily good at processing it; we have seemingly specialized brain regions for it; and music is found virtually everywhere, with certain fundamentally similar characteristics.

To my mind, Pinker's arguments that we are not the universal-learning machines we are often believed to be (something he has argued in all his books) are convincing. And his arguments that language possesses all the hallmarks of design (and analogous arguments by others in the case of music) are highly persuasive.

Language and music, on the one hand, and the human brain, on the other, are designed to fit one another.

But there is a gnawing problem, one Pinker himself implicitly reveals on the first page of his book, in the passage I quoted above: the octopus, club soda, and soap opera excerpts were *written*. My ability to comprehend Pinker's examples—and all his books, and, well, everything I have ever come to know and admire about him—relied on writing and reading.

Why is reading a problem for the notion of language and music instincts? Because, like language and music, our ability to read *also* has the hallmarks of design . . . and yet we *know* we have no reading instinct.

We know there's no reading instinct because writing is too recent, having been invented only several thousand years ago; in fact, it didn't take hold among a large fraction of the population until just a few generations ago. There's a good chance most of your great-great-great-grandparents didn't read.

And yet, despite the fact that we cannot possibly have specialized reading mechanisms in our brains, reading has the same appearance of instinct, much like language and music. Reading is astoundingly complex—to this day, we cannot build effective handwriting-recognition machines—and yet we display machinelike proficiency at reading. Children learn to read at about twice the age at which they can comprehend speech, but when they do learn, their reading experiences are meager compared to those for speech, To put it in context, they're often reading before they're competent at pouring milk into cereal, wiping their bottoms, or even engaging in stereotypical ape behaviors like turning somersaults and climbing monkey bars. Once we've learned, we read automatically and effortlessly, and reading is arguably more pervasive in our lives today than speech. Our brilliantly capable reading brains even appear to have regions specialized for reading (one is called the "visual word form area"), which researchers like the neuroscientists Stanislas Dehaene and Laurent Cohen discuss, and which Dehaene takes up in his recent book, *Reading in the Brain*. The whiff of a reading instinct is also apparent in the near universality of writing and reading. Writing is found in nearly every human society today, and there are strong universal tendencies across writing (e.g., in the number of strokes per character among phonemic writing systems like ours, and in the ways that strokes can interconnect to build characters, something I discussed in Chapter 4 of *The Vision Revolution*).

If we can appear to have a reading instinct without actually having one, perhaps the appearance of instincts for language and music is an illusion, too. Perhaps the story of the origins of speech and music is the same as the story underlying our ability to read, whatever that story might be.

It does not escape Pinker's notice that his illustration of language's power is communicated to a reader, not a listener. He says in the paragraph following the octopus-soda-soap excerpts:

> True, my demonstrations depended on our ability to read and write, and this makes our communication even more impressive by bridging gaps of time, space, and acquaintanceship. But writing is clearly an optional accessory; the real engine of verbal communication is the spoken language we acquire as children.

Writing is optional, Pinker says, but optional for what? Speech and writing serve distinctly different functions. As Pinker notes, writing, but not speech, can bridge space and time, giving writing a power akin to a superpower (for example, if I'm dead when you're reading this, then you're not merely reading, but spirit reading!). And as I discuss in *The Vision Revolution*, writing serves functions that audio recordings (which also bridge space and time) cannot, allowing the reader to interact with other minds and upload content so efficiently that it changed us from *Homo sapiens* to a universally programmable *Homo turingipithecus*. And these distinctive functions of writing are not optional: recorded history and modern civilization depend on it!

At any rate, optional or not, we appear to be designed to read, and yet we have no reading instinct. How is this possible?

The answer is that, rather than our brains being designed for reading, reading is designed for our brains. Writing is a technology that has been optimized over time by the forces of cultural selection to be good for our visual system. We have no reading instinct. Instead, writing has a brain instinct (i.e., is designed for the brain), something neuroscientist Stanislas Dehaene calls "neuronal recycling."

In my research and in my previous book, *The Vision Revolution*, I provided evidence to support a specific theory of how culture managed to shape writing for the brain: writing was culturally selected to look, in fundamental respects, *like nature*, which is the look our evolutionarily-illiterate visual system is highly competent at processing. Writing doesn't have a brain instinct so much as a *nature* instinct.

In the case of writing, then, instinct is not responsible for the appearance of design. The designer is not natural selection, but cultural selection. The tight fit between reading and the brain is because reading has been bent to the brain, not the other way around. And the tight fit was achieved via what I call *Nature-Harnessing*: mimicking nature so as to harness evolutionarily ancient brain mechanisms for a new purpose.

And now we are poised to see the purpose of *this* book.

If cultural selection can give us writing shaped like nature that is thereby optimized for our visual system, and can do so in just several thousand years, then imagine how well optimized for our brains speech and music may be if they have been culturally evolving for *hundreds* of thousands of years to be good for our auditory systems! What if writing, speech, and music are *all* products of culture, but—consistent with the fact that we're not general-purpose machines—they are highly designed technologies shaped for our minds?

And, more specifically, what if, just as writing *looks* like nature, these two auditory capabilities—speech and music—have come to *sound* like nature, and thereby to harness ancient, highly efficient brain mechanisms that were never intended for language or music?

But what in nature might speech and music sound like? *That's* the topic of the book. Before getting into speech and music, however, we must discuss in more detail the general nature-harnessing approach that I believe explains writing, speech, and music—and explains who we are today. And that's the topic of Chapter 1.

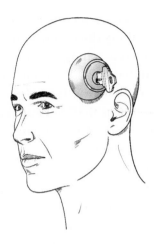

# Nature-Harnessing

## DEEP SECRETS

It isn't nice to tell secrets in front of others. I recently had to teach this rule to my six-year-old daughter, who, in the presence of other people, would demand that I bend down and hear a whispered message. To my surprise, she was genuinely perplexed about why communicating a message only to me, in the presence of others, could possibly be a bad thing.

Upon thinking about it, I began wondering: What exactly *is* so bad about telling secrets? There are circumstances in which telling secrets would appear to be the appropriate thing to do. For example, if Dick and Jane are over for a formal dinner at my house, and in order to spare my wife some embarrassment I lean over and whisper "Honey, there's chocolate on your forehead," is that wrong? Wouldn't it be worse to say nothing, or to say it out loud (or in a published book)?

The problem with telling secrets is not that there aren't things worth telling others discreetly. Rather, the problem is that when we see someone telling a secret, it taps into a little program in our head that goes, "That must be very important information—possibly about *me*. Why else keep it from me?" The problem with secrets is that we're all a bit paranoid, and afflicted with a bad case of "Me, me, me!" The result is that secrets get imbued with weighty importance, when they mostly concern such sundries as foreheads and sweets.

Not only do we tend to go cuckoo over the covert, attributing importance to unimportant secrets, but we also have the predisposition to see secrets where there are none at all. Our propensity to see nonexistent secrets has engendered some of the most enduring human preoccupations: mysticism and the occult. For example, astrology, palm reading, and numerology are "founded" upon supposed secret meanings encoded in the patterns found in stars, handprints, and numbers, respectively. Practitioners of mysticism believe themselves to be the deepest people on Earth, dedicated to the ancient secrets of the universe: God, life, death, happiness, soul, character, and so on. Astrological horoscopes don't predict the morning commute, and palm readers never say with eerie omniscience, "Don't eat the yogurt in your fridge. It's moldy!" The secrets are not only deep, but often personal: they're about ourselves and our role in the universe, playing on our need for more "me, me, me."

Even those who pride themselves on not believing in mystical gobbledygook often still enjoy a good dose of deep ancient secrets in fiction, which is why Dan Brown's *The Da Vinci Code* did so well. In that novel secret codes revealed secret codes about other secret codes, and they all held a meaning so deep that people were willing to slay and self-flay for it.

Secrets excite us. But stars, palms, and numbers hide no deep, ancient secrets. (Or at any rate, not the sort mystics are searching for.) And stories like the *The Da Vinci Code* are, well, just stories. What a shame our real world can't be as romantic as Dan Brown's fictional one, or the equally fictional one mystics believe they live in. Bummer.

But what if there *are* deep and ancient secrets? Real ones, not gobbledygook? And what if these secrets *are* about you and your place in the universe? What if mystics and fiction readers have been looking for deep secrets in all the wrong places?

That's where this book enters the story. Have *I* got some deep secrets for you! And as you will see, these ancient secrets are much closer than you may have thought; they're hiding in plain sight. In fact, as I write these very words I am making use of three of the deepest, ancientest secret codes there are. What are these secrets? Let me give you a hint. They concern the three activities I'm engaged in right now: I am *reading* (my own writing), *listening to speech* (an episode of bad TV to keep me awake at 2 a.m.), and *listening to music* (the melodramatic score of the TV show). My ability to do these three things relies on a code so secret few have even realized there's a code at all.

"Code, schmode!" you might respond. "How lame is that, Changizi? You tantalize me with *deep* secrets, and yet all you give me are run-of-the-mill writing, speech, and music! Where are the ancient scrolls, Holy Grails, secret passwords, and forgotten alchemy recipes?"

Ah, but . . . I respond. The secrets underlying writing, speech, and music *are* immensely deep ones. *These secret codes are so powerful they can turn apes into humans. As a matter of fact, they **did** turn apes into humans.* That's deeper than anything any mystic ever told you!

And it is also almost certainly truer. So shove that newspaper horoscope and that Dan Brown novel off your coffee table, and make room for this nonfiction story about the deepest ancient secret codes we know of . . . the ones that created us.

To help get us started, in the following section I will give you a hint about the nature of these codes. As we will see, the secret behind the codes is . . . *nature itself.*

## MOTHER NATURE'S CODE

If one of our last nonspeaking ancestors were found frozen in a glacier and revived, we imagine that he would find our world jarringly alien. His brain was built for nature, not for the freak-of-nature modern landscape we humans inhabit. The concrete, the cars, the clothes, the constant jabbering— it's enough to make a hominid jump into the nearest freezer and hope to be reawakened after the apocalypse.

But would modernity really seem so frightening to our guest? Although cities and savannas would appear to have little in common, might there actually be deep similarities? Could civilization have retained vestiges of nature, easing our ancestor's transition? And if so, why *should* it—why would civilization care about being a hospitable host to the freshly thawed really-really-great-uncle?

The answer is that, although we were born into civilization rather than melted into it, from an evolutionary point of view we're an uncivilized beast dropped into cultured society. We prefer nature as much as the next hominid, in the sense that our brains work best when their computationally sophisticated mechanisms can be applied as evolution intended. Living in modern civilization is *not* what our bodies and brains were selected to be good at.

Perhaps, then, civilization shaped itself for *us*, not for thawed-out time travelers. Perhaps civilization possesses signature features of nature in order to squeeze every drop of evolution's genius out of our brains for use in the modern world. Perhaps we're hospitable to our ancestor because we have been hospitable to *ourselves*.

*Does* civilization mimic nature? I believe so. And I won't merely suggest that civilization mimics nature by, for example, planting trees along the boulevards. Rather, I will make the case that some of the most fundamental pillars of humanity are thoroughly infused with signs of the ancestral world . . .

and that, without this infusion of nature, the pillars would crumble, leaving us as very smart hominids (or "apes," as I say at times), but something considerably less than the humans we take ourselves to be today.

In particular, those fundamental pillars of humankind are (spoken) language and music. Language is at the heart of what makes us apes so special, and music is one of the principal examples of our uniquely human artistic side.

As you will see, the fact that speech and music *sound like other aspects of the natural world* is crucial to the story about how we apes got language and music. Speech and music culturally evolved over time to be simulacra of nature. Now *that's* a deep, ancient secret, one that has remained hidden despite language and music being in right front of our eyes and ears, and being obsessively studied by generations of scientists. And like any great secret code, it has great power—it is so powerful it turned clever apes into Earth-conquering humans. By mimicking nature, language and music could be effortlessly absorbed by our ancient brains, which did *not* evolve to process language and music. In this way, culture figured out how to trick nonlinguistic, nonmusical ape brains into becoming master communicators and music connoisseurs.

One consequence of this secret is that the brain of the long-lost, illiterate and unmusical ancestor we unthaw is no different in its fundamental design from yours or mine. Our thawed ancestor might do just fine here, because our language and music would harness *his* brain as well. Rather than jumping into a freezer, our long-lost relative might instead choose to enter engineering school and invent the next-generation refrigerator.

The origins of language and music may be attributable, not to brains having evolved language and music instincts, but rather to language and music having culturally evolved *brain instincts*. Language and music shaped themselves over many thousands of years to be tailored to our brains, and because

our brains were cut for nature, language and music mimicked
nature . . . and transformed ape to man.

## UNDER THE RADAR

If language and music mimic nature, why isn't this obvious to
everyone? Why should this have remained a secret? It's not as
if we have no idea what nature is like. We're not living on the
International Space Station, and even those who are on the
Space Station weren't *raised* up there! We know what nature
looks and sounds like, having seen and heard countless examples of it. So, given our abundant experiences of nature, why
haven't we noticed the signature of nature written (I propose)
all over language and music?

  The answer is that, ironically, our experiences with nature
don't help us consciously comprehend what nature *in fact* looks
and sounds like. What we are aware of is already an assembled
*interpretation* of the actual data our senses and brains process.
This is true of you whether you are a couch potato extraordinaire or a grizzled expedition guide just returned from Mada-
gascar and leaving in the morning for Tasmania.

  For example, I am currently in a coffee shop—a setting
you'll hear about again and again—and when I look up from
the piece of paper I'm writing on, I see people, tables, mugs,
and chairs. That is, I am consciously aware of seeing these
*objects*. But my brain sees much more than just the objects. My
early visual system (involved in the first array of visual computations performed on the visual input from the retina) sees
the individual contours, and does *not* see the combinations of
contours. My *intermediate-level* visual areas see simple combinations of several contours—for instance, object corners such as
"L" or "Y" junctions—but don't see the contours, and don't
see the objects. It is my *highest-level* visual areas that see the
objects themselves, and I am conscious of my perception of

these objects. My conscious self is, however, rarely aware of the lower hierarchical levels of visual structure.

For example, do you recall the figure at the start of the chapter—the person's head with a lock and key on it? Notice that you recall it in terms referring to the *objects*—in fact, I just referred to the image using the terms *person, head, lock* and *key*. If, instead, I were to ask you if you recall seeing the figure that had a half dozen "T" junctions and several "L" junctions, you would likely not know what I was talking about. And if I were to ask you if you recall the figure that had about 40 contours, and I then went on and described the geometry of each contour individually, you would likely avoid me at cocktail parties.

Not only do you (your conscious self) not see the lower-level visual structures in the image, you probably won't find it easy to talk or think about them. Unless you have studied computational vision (i.e., studied how to build machines that see) or are a vision scientist, you probably haven't thought about how contours intersect one another in images. "Not only did I not see T or L junctions in the image," you might respond, "I don't even know what you're talking about!" We also have great trouble talking about the orientation and shapes of contours in our view of three-dimensional scenes (something that came to the fore in the theory of illusions I discussed in *The Vision Revolution*).

Thus, we may *think* we know what a chair looks like, but in a more extended sense, we have little idea, especially about all those lower-level features. And although parts of our brain *do* know what a chair looks like at these lower levels, they're not given a mouthpiece into our conscious internal speech stream. It is our inability to truly grasp what the lower-level visual features are in images that explains why most of us are hopeless at drawing what we see. Most of us must undergo training to become better at accessing the lower levels, and even some of the great master painters (such as Jan Van Eyck) may have projected images onto their canvases and *traced* the lower-level structures.

Not only do we not truly know what nature looks like, we also don't know what it *sounds* like. When we hear sounds, we hear the meaningful events, not the lower-level auditory constituents out of which they are built. I just heard someone at the next table cutting something with her fork on a ceramic plate. I did not consciously hear the low-level acoustic structure underlying the sound, but my lower-level auditory areas *did* hear just that.

For both vision and audition, then, we have a hierarchy of distinct neural regions, each a homunculus ("little man") great at processing nature at *its* level of detail. If you could go out for drinks with these homunculi, they'd tell you all about what nature is like at lower and middle hierarchical scales. But they're not much for conversation, and so you are left in the dark, having good conscious access only to the final, highest parts of the hierarchy. You see objects and hear events, but you do not see or hear the constituents out of which they are built (see Figure 2).

You may now be starting to see how language and music could mimic nature, yet we could be unaware of it. In particular: what if language and music mimic all the lower- and middle-level structures of nature, and only fail to mimic nature at the highest levels? All our servant homunculi would be happily and efficiently processing stimuli that appear to them to be part of nature. And yet, because the stimuli may have a structure that is not "natural" at the highest hierarchical level, our conscious self will only see the *dissimilarity* between our cultural artifacts and nature.

Why should we believe what we can't consciously perceive—that language and music mimic nature at all but the highest hierarchical level? Why not go all the way and make language and music *completely* like nature?

Let's not forget that language and music are not *merely* trying to mimic nature. They have *jobs* to do: writing is for putting thoughts on the record, speech is for transmitting thoughts to others, and music is perhaps for something like evoking feelings in others. Language and music want to capture as much of the structure of nature as they can so that they have an easy ride into our brains, but they must serve their purpose, and will have to sacrifice nature-mimicry when it is necessary to do so.

So one can see how sacrifices of nature-mimicry may sometimes be part of doing business. But why should the sacrifices be up near the top, where we have greater conscious access? The principal reason for this is that if the earlier regions of the hierarchy receive stimuli that *they* can't make any sense of, then they will output garbage to the next higher level, and so all levels above the unhappy level will be unhappy. Breaking nature-mimicry at one level will break it at all higher levels

For example, I have argued in earlier research and in *The Vision Revolution* that writing looks like nature. In particular, I have suggested that written words look like visual objects. But words do not necessarily look natural at *all* levels up the hierarchy. Strokes look like contours, and letters look like object junctions; and thus the lower and middle levels of your visual hierarchy are happy. But because in alphabetic writing systems the letters in a word depend on how it is spoken, there is no effective way to make entire words look like objects. (For example, the junction-like letters in the words you are currently reading are simply placed side by side, which is not the way junctions in scenes are spatially related.) Your highest-level regions, of which you are most directly aware, only notice the nonnatural look of written words. And when visual signs do more closely match the visual structure of objects at the highest levels, people *do* see the resemblance to nature—this is why trademark logos and logographic writing systems like Chinese look (to your conscious self) much more object-like than the words you're reading here.

My claim in this book that language and music mimic nature must be understood in this light. I claim that they mimic nature, indeed, but not necessarily "all the way up." The reason why writing, speech, and music don't *obviously* seem like nature is that nature is *not* being injected at the higher levels, perhaps—as we've seen with writing—in order to better accomplish the functions they are designed to carry out.

We see, then, why it is that the nature-mimicry in language and music has remained a secret for so many millennia. If only your lower-level visual and auditory areas could speak! They'd have long ago let you know that language and music are built like nature. Because those lower homunculi are *part* of you, there is a sense in which you have known about this ancient, deep secret code all along. Pieces of meat inside you knew the secret, but weren't telling. In this light, one can view this book as a kind of psychoanalysis—if you're into that— digging up the homunculus-knowledge you already have deep inside you, and working through the ways it shaped who you are today.

## NATURE'S HARD CORE

"Language and music mimic nature."

I *will* try to convince you of that over the course of this book. But we humans are very bad judges of whether a stimulus is natural or not, as we just finished discussing. How, then, are we to have any idea whether language and music do or do not mimic nature? And isn't the phrase "mimicking nature" awfully imprecise?

Indeed, it *is* imprecise. Part of the book's point is to make this more precise—to say *specifically which* aspects of nature are mimicked by language and music. But can this really be done? How can we possibly know the details of the natural world our ancestors experienced? It is, after all, *that* version of nature that our brains evolved to be good at processing, and that was a long time ago. In order to show that language and music mimic the primordial natural habitats that shaped our brains, it would seem that we have no choice but to start familiarizing ourselves with the sights and sounds of the savanna, and wherever else our ancestors hung out. .

A field trip is in order. Like all natural habitats, savannas feature a hodgepodge of characters and settings, and we must take a visual and auditory inventory. We will need images from all the savanna's habitats, including acacia trees, tall grass, sunsets, rocks, giraffes, lions' manes, and termite mounds. And we will need to record the sounds within those habitats, including wind, rustling leaves, bird calls, insect buzzes, and rhino grunts. Once we have built an encyclopedia of how our ancestral world looked and sounded, we can *then* ask whether language and music mimic nature.

Hold it right there! Sorry, but I'm pulling your leg. That is definitely *not* what we're going to do. Trekking across the Serengeti with a camera and tape recorder sounds grand, but it isn't the route to a compelling scientific explanation. We want *generalizations* about what visual nature looked and sounded like for our ancestors, not merely an itemized list of all the furniture in the savanna. We need to grasp the fundamental "grammar" of nature. We need to pick apart nature, carve it at its joints, and elegantly summarize its structure. With such a "grammar" in hand, we will be able to make powerful predictions about what a nature-mimicker should be like . . . and thus, what language and music should be like.

Might there be fundamental regularities that cut across a wide swath of terrestrial habitats? Could it be that, although there are large salient differences across habitats, there are nevertheless deeper respects in which they are all similar? Although the savanna and any other specific habitats shaping our ancestors have hosts of peculiar features, might the sights and sounds of these habitats nevertheless have the same fundamental "core grammar"?

If there *is* a "core grammar" to nature, then this core would have been a highly steady invariant over our evolutionary history, and would thus have been a strong shaper of our visual and auditory systems. Sensory structure specific to the savanna or other particular habitats would, on the other hand, have been highly variable and intermittent over evolutionary time, and consequently less important for understanding what our visual and auditory systems became good at.

It's this hard core of nature that we want. But is there one? Yes, indeed. There *are* solid core grammars underlying the structure of visual and auditory nature—there are "universals" in the structure of nature. That's what I'll endeavor to show you in this book; and *then* I'll show you that language and music mimic these cores. I'll give you a preview of these hard cores in the next section, when I introduce the central tenets of the theory—that is, when I first reveal the big secrets.

## THE REVEAL

At this point, we have discussed the possibility that language's and music's mimicry of nature could be what enabled us humans to acquire language and music. We also took up why, if this were so, it would not be obvious to us. And in the section just above, I made myself clearer about the role that "nature" will play in understanding the origins of language and music: the goal is to find fundamental principles underlying the structure of nature (as opposed to a catalog of savanna paraphernalia), so that we'll be in a strong position to ask whether these fundamental principles underlie language and music.

What I have *not* yet done is give you any specifics on what the hard cores in nature *are* that are being mimicked by language and music. That is, I haven't revealed what any of the "ancient secrets" might actually be. Let's rectify that, and simultaneously summarize the book's two main theses, concerning speech and music.

Chapter 2 of the book is about the secret code underlying speech. Here's the secret: *human speech sounds like solid-object physical events*. Notice that this secret makes no mention of rustling leaves or rhino grunts. Instead, the "nature" that speech mimics encompasses a very broad class of events, namely those among solid objects. The main observation is that events involving solid objects bumping and crashing about have a signature core auditory structure, and I will provide evidence that human speech has this signature structure. Speech can thereby get into our brains by harnessing auditory-recognition mechanisms we have long possessed for processing the "pinball" sounds of nature. This secret code gives us hominids the power to recognize speech without having speech-recognition mechanisms.

And here is the deep secret underlying music, which is the topic of Chapters 3, 4, and 6: *music sounds like humans moving and behaving (usually expressively)*. Notice how general the notion of "nature" is here. It isn't the sounds of people's heartbeats, or heavy breathing, or missionary-style sex, or skipping—it is the sounds of humans behaving. When we carry out behaviors we tend to make noise, and our auditory system can infer a lot about each other's behavior from the noise. Music has the signature auditory structure of humans doing stuff, and can thus get into our brains by tapping into our auditory recognition mechanisms for identifying the actions of other people. No music-processing mechanisms are required. This secret code of sounding like expressive human movers is what allows music to flow into our auditory system and be understood by it, even though we possess no special brain gears for processing music.

Although it is the ancient secret codes lying beneath speech and music that I'll be whispering about in this book, this is not the first time I have written about deep secrets of this kind. In my previous book, *The Vision Revolution*, I wrote about (among other topics) another ancient secret, the one explaining how we hominids came to have a *written* history. And the secret is this: *writing looks like three-dimensional scenes with opaque objects.* Writing looks like nature, but as before, nature of a very general kind—no images of acacia trees or termite mounds are needed. Writing has come to mimic the contour combinations occurring in natural scenes with opaque objects, and in such a way that written words mimic the structure of visual objects. Writing gets into our brains by harnessing our visual object-recognition mechanisms. The secret code of looking like nature is what allows writing to be read by us hominids without any reading mechanisms in our brains.

The "nature" stories of the origins of speech, music, and writing are, then, not in the least about acacia trees or the other particulars in the rummage shop of our ancestors, but rather about solid-object physical events, human movement sounds, and opaque objects in a three-dimensional world. "Nature" is a highly general notion, just what is needed to make theoretical headway and empirical testing possible and practical.

And although the notions of "nature" I will rely on are very general, they are not so general that they include everything. For example, "solid-object physical events" covers a wide swath, but it doesn't cover sounds made by air and water. And "opaque objects in a three-dimensional environment" is fundamental, but a habitat with semitransparent objects (like clouds at high altitudes) would not be included.

Distinguishing between the surface features of habitats (which vary wildly from habitat to habitat) and the core features (found in most or all habitats) is helpful in understanding why, even if writing, speech, and music have underlying core similarities, they nevertheless come in such tremendous variety. If nature were all core—if it had little or no variability across habitats—then our visual and auditory systems would have evolved to be competent at processing just the very specific kinds of stimuli in the world. Language and music that harnessed such a brain would be expected to have a

NATURE-HARNESSING                    21

very specific and consistent surface structure, something they do not, in fact, have. If, instead, as is the case, there is a small core of invariant structure to nature, yet loads of variability across habitats, one would expect us to end up with brains that are more open-minded about what they're willing to accept. Our brains would be most competent at processing stimuli that have the core signature, but in other respects, our brains would be open to many variants. Language and music that harnessed this kind of open-minded brain would be expected to take widely varying shapes across cultures, but to share certain similarities. This is a much more accurate description of language and music as found on Earth: subject to large differences across cultures, but sharing certain core structural characteristics across cultures.

You have now gotten a peek at the ancient secret codes hidden inside speech and music. Hopefully you can appreciate their generality, and appreciate why it might be that the natural structure in speech and music has stayed hidden from us. In the next and final section of this chapter, I will be as clear as I can about how my Nature-Harnessing theory differs from other stances on the origins of language and music, and I will justify why we should *expect* language and music have nature instincts (i.e., designed to mimic nature) rather than just brain instincts (i.e., designed to be well-shaped for the brain).

## PURPS VS. QUIRKS

In the Introduction, I touched upon two standard, contrasting viewpoints on origins, the first being that we evolved brains specialized for language and music (i.e., we have instincts for these things), and the second that, on the contrary, we evolved to be general-purpose, universal learning machines that handle these artifacts because we can learn *lots* of unusual stuff. I suggested that language and music seem unlikely to be instincts because writing, too, reeks of instinct, but is definitely not an instinct. But I also intimated that there is a wealth of data and argument—summarized and argued convincingly in Pinker's books, for example—that we do not possess blank-slate brains. How, then, are *predisposed* brains like ours able to learn any human language and comprehend music—among the most complex and sophisticated computational tasks on Earth—if we're neither designed specifically for it nor particularly impressive general learners?

The answer is that once culture got up and running, there was a new blind watchmaker in town. Cultural evolution could, over comparatively short periods of time, fit language and music into the shapes our grooved (non–blank-slate) brains are able to learn. It is not so much that our brains learn language and music, but rather that culture learned how to package language and music so that they fit right into our brains. Culture learned how to harness us.

Despite the title of this book, there is nothing new about the idea that we are harnessed by culture, that cultural artifacts may have been selected to be structured well for our brains. What *is* new here is that I am putting forth specific proposals for how culture actually goes about harnessing us. Saying that language and music might be shaped for the brain doesn't take us very far in understanding the shape of language and music, because we don't have a good understanding of the brain. What we need is a *general theory* of harnessing. And *Nature-Harnessing* is the theory I am proposing.

Earlier I said that language and music have evolved to possess a brain instinct, rather than the brain having evolved to possess language and music instincts. But in a sense, in this book I am arguing that language and music have, not a brain instinct, but a *nature* instinct. Language and music carry in them the structure not of the brain so much as of nature, which of course is just right for the brain—because the brain is just right for nature (see Figure 1).

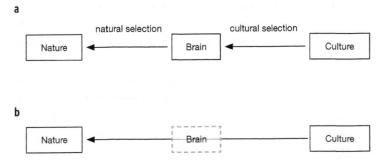

**FIGURE 1.** **(a)** The brain was shaped by natural selection for nature, and culture was shaped by cultural selection for the brain. **(b)** By shaping culture to look like nature, culture will tend to end up shaped well for the brain. And, importantly, we scientists can hope to get a handle on this without having to understand the detailed brain mechanisms. The arrow cutting through the brain and going from culture to nature is meant to symbolize my nature-harnessing theoretical approach which drives this book. It means that my theory will pretend there is a single arrow like this, where culture has been selected to be shaped like nature. This is a simplification of the more detailed picture in **(a)**, and the greater simplicity is a boon to a scientist because the most complicated object in the universe—the brain—has been removed from the "equation".

We have to be more careful, however, because brains optimized for nature can sometimes like nonnatural things as well. Our mechanisms have been selected for because they work very well on the inputs our ancestors would have experienced. When those natural stimuli are the input, our mechanisms work as they are supposed to—it's their purpose (or "purp").

But those same mechanisms don't typically just sit quietly when nonnatural stimuli are inputted into them. They do *something*. And what they do depends entirely on the implementation details of the mechanism. Because the mechanism wasn't designed to handle that kind of input, who knows what the mechanism might do? Mechanisms have *quirks*. For example, it is presumably a quirk that certain flashing lights have a propensity to induce seizure.

Brains were selected for their purps, but they end up with lots of quirks as well. When language and music culturally evolved to be structured for our brains, it didn't matter whether it was the purps or the quirks that were harnessed, so long as the process worked. But if language and music actually came to harness our quirks more than our purps, then the strategy that culture uses would not be nature-harnessing so much as *quirk*-harnessing. And if *that* were the case, I wouldn't have much of a book left! That is, in this book I am claiming that the principal strategy culture used to harness our brains for language and music is not quirk-harnessing, but purp-harnessing . . . and that culture did its purp-harnessing by mimicking nature, just the thing to ensure that our brain mechanisms run as "purposely" designed.

So, is harnessing about the purps or the quirks? Does culture harness the brain by looking and sounding like nature and thus making the brain function as intended, or does it harness the brain by shaping itself in a way that elicits the brain to function in some quirky accidental manner? Because, as I just said, cultural evolution doesn't care what it harnesses so long as it works, both purps and quirks are surely both part of the full story of how language and music fit themselves to us. There's no reason, then, to expect that the quirks should completely *dominate* the story of harnessing. And if that's the case, then there's a role for the purps, and for nature-harnessing. Whew!

Actually, I can say more than just that nature-harnessing is unlikely to be completely useless for understanding harnessing. On the contrary, I expect nature-harnessing to be the *key* to how cultural evolution harnessed us, and quirks to be just a small side story. There are two reasons why I don't think the quirks are the main driver. The first reason is that quirks are not smart enough, and the second reason is that *I* am not smart enough.

Stupid quirks first. If I were to open up the "V" of a stapler, hold one end in my hand, and try to hit you with the swinging end, then I would have created a hitting device (and lost a reader). I would thereby have harnessed the stapler for a new function. But I would have harnessed a *quirk* of the stapler, not a purp. Staplers are not designed to be weapons, or to be swung around like that, at all. They are, accordingly, unlikely to be any good at it; at best, they'll be *nowhere near* as efficient as tools designed for hitting. My stapler hitting device is, in essence, the worst pair of nunchucks ever devised. You don't get powerful functionality by accident. If, instead, I were to use the stapler to fasten a pile of leaves together, that would be a case where I have harnessed the *purp*. Staplers may not be for stapling leaves, but leaves clearly resemble paper (in the respects relevant for staplers), which is just what staplers *were* designed for. So, the first reason why quirk-harnessing will be a minimal part of the story of harnessing is that cultural selection will favor the bits of us that are highly engineered masterpieces, not accidental side effects.

Quirks may be stupid, but cultural evolution may sometimes tap into them anyway. After all, who hasn't tried to remove a staple with a pen tip, or tried to bang a nail in with the handle of a screwdriver? And this leads naturally to the second problem with quirks, which is that *I'm* not smart enough to figure them out. First, there's no general characterization of the quirks. A quirk occurs whenever the brain is confronted with a non-natural stimulus, and although there may be a "hard core" for natural stimuli, there are no core ways of being *un*natural. For example, pens can be used for stabbing, picking your teeth, scratching an itch, eyeliner, penny flicking, donut-hole making . . . clearly this list has no end. But the list of what pens are *for* is short: pens are for writing on paper.

And not only are there piles of quirky ways to use a mechanism, but there will typically be no simple characterization of how the mechanism will react in any specific case. Whereas the proper function of a pen can be activated by a mechanism characterized by a description something like this—"a hand holding the pen and lightly moving on the surface of the paper, leaving ink"—the mechanistic descriptions for

different quirks will tend to differ wildly, and to refer to physical aspects of the pen that are not part of any description of
writing. For example, good penny flicking depends on a pen's
rigidity being in the right range. And the pen I'm holding
right now could serve as a container for sand, which depends
on how the pen fits together so that it has room left over on
the inside. These and many other peculiar characteristics of
the mechanism aren't relevant for understanding the proper
function of the pen. And when it comes to the brain, we are
woefully ignorant of its mechanisms, and so it is immensely
difficult to determine which characteristics are central to its
natural operation and which are not. The quirks are difficult
to comprehend, but the purps are comparatively simple. I
have a *hope* of wrapping my head around the fundamental
core regularities found in nature and characterizing the
brain's likely response (the purps), but practically no hope of
doing so for the quirks.

To sum up, there's no reason to believe that harnessing is
completely dominated by the quirks. On the contrary, because
most quirks are not truly useful for anything, whereas focused
usefulness is the very essence of purps, purps are far more
likely to be harnessed. There will inevitably be some facets of
language and music that are not mimicking nature, but are,
rather, shaping themselves to fit the quirks. But in this book
I'll ignore these quirks, for the reasons I just went over. To the
extent that language and music have come to harness quirks
despite their deficiencies, I'll leave that to future scientists to
unravel, because it is far above my pay grade.

Now, with quirks out of the way, the fundamental argument
structure of nature-harnessing can be illustrated by Figure 1b.
If the brain in the story "from nature to brain to culture" is
covered over. that leaves only nature and culture, highlighting
the hypothesis that culture mimics nature.

Figure 2 below shows the three cases of nature-harnessing
I have examined in my research: writing, speech, and music.

It shows the mechanisms in the brain each harnesses, and also the natural stimuli the brain mechanisms were selected to process. Writing was covered in *The Vision Revolution*. The other two rows in Figure 2 are for speech and music, the cultural artifacts taken up in *this* book, with Nature-Harnessing as the overarching theme.

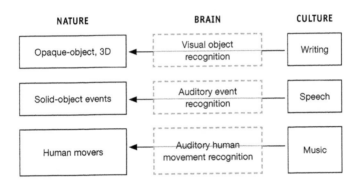

**FIGURE 2.** The structure of the book's argument. For example, for the first row, writing shaped itself (via cultural selection) for our visual object recognition mechanisms in the brain, and these mechanisms were, in turn, shaped (via natural selection) for recognizing three-dimensional scenes with opaque objects strewn about. Supposing that writing shaped itself mostly for the brain's "purps" and not the quirks, then writing is expected to principally shape itself to look like three-dimensional scenes with opaque objects. The next two rows are the main topics of *this* book.

And now we're ready for the meat of this book. In Chapter 2, I describe how speech sounds like solid-object physical events, and in Chapters 3, 4, and the Encore (at the end of the book), I describe how music sounds like people moving. In the fourth chapter of my previous book, *The Vision Revolution*, I described how writing looks like 3-D scenes with opaque objects. With these three cases made, the conclusion I would like the reader to draw is that Nature-Harnessing—not instinct, and not a general-purpose brain—is the general mechanism by which we came to have these powers.

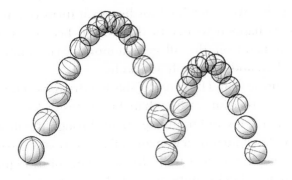

# CHAPTER 2

# Speech Events

## GRASSHOPPER

In M. Night Shyamalan's movie *The Village*, a young woman, Ivy, sets off on a journey into an unknown forest. She has persuaded the elders of her tribe to let her find other people on the far side of the forest, get medicine, and return to save the life of her sick lover. She has no knowledge of anything beyond the several acres of her village, except that beyond their meadow and inside the forest are chilling, otherworldly beasts which occasionally invade the village and carve up one of the pets.

As if this quest were not harrowing enough, there's an important fact I left out: she is blind. Now, the village leaders know the truth about what's beyond their meadow—no beasts (but the costumed elders themselves), just woods, and then modern civilization, from which they've sheltered their children. That's why they allow her to go into the forest. But no one of Ivy's generation knows this. And neither do we, the moviegoers. We're terrified for her. As it turns out, terrifying things *do* happen to her in that forest, because a monster (really a man from the village in a monster costume) secretly follows her, and eventually attacks her.

The movie would be considerably less dramatic if our female heroine were deaf, rather than blind. Instead of a woman waving her arms and tramping about through the thorny tangles, we'd be watching a woman walking normally through the forest, keeping to deer trails. In fact, many of us regularly do just this, wearing headphones and blasting music as we deafly, yet deftly, jog through our local park. This would not quite elicit the thrill Shyamalan had in mind. A deaf person on a forest quest does not make a good movie. Being deaf just doesn't seem like much of a big deal compared to blindness. If not for the inability to hear speech, we might hardly miss our auditory systems if they fell out through our ears.

Then again, there's another twist to the story that may change one's feeling about audition: our young blind heroine *defeats* her attacker. She kills him, in fact. She may look out of sorts crashing into trees, but her hearing makes it impossible for her attacker to sneak up on her. Especially in the forest. Had she been deaf, not blind, her attacker could have whistled "Dixie" with an accordion accompaniment while following her through the woods and still taken her completely by surprise.

If deaf-maiden-alone-in-the-forest is not spine-tingling to movie audiences, it is only because we tend not to appreciate all that our ears do for us beyond language. Providing a sneak-proof alert system is just one of the many powers of audition.

The greatest respect for our ears is found among blind kung fu masters. Every "Grasshopper" learns from his old blind master that by attending to and dissecting the ambient sounds around oneself, it is possible to sense how many attackers surround one, their locations, stances, weapons, intent, confidence level, and which one is the enemy mastermind. I once saw, in an old movie, one of these scrawny geezers defeat six men using only a baseball bat wielded upside down. But you don't have to be a fictional blind kung fu master to have a mastery of audition and know how to sense the world with it. We all do; we just don't get all "Grasshopper" about it. Our brains have a mastery of it even if we've never thought about it.

In fact, when I first began pondering whether speech might sound like natural events, I had great difficulty thinking of *any* important natural-event sounds. I was initially dumbfounded: what is so useful about having ears that nearly all vertebrates have them? It seemed to me that *I* primarily use my ears for listening to speech, and *that* obviously cannot explain why all those other vertebrates have ears as well. Sure, it is difficult to sneak up on me, but one hardly needs such a fine-tuned ear and auditory system for a simple alarm.

After some months of contemplation, however, I came to consciously appreciate my ability to use sound to recognize the world and what's happening around me. I began to notice every tap, clink, rub, burble, and skid. And I noticed how difficult it was for me to do anything without making a sound that gave away what I was doing, like eating from my daughter's Halloween stash. When you're next at home and your family is active around you, close your eyes and listen. You will hear sounds such as the plink of a spoon in a coffee mug, the scrape of a drawer opening, or the scratch of crayons on drywall. It will typically take some time before you hear an event that you *cannot* recognize. In the late 1980s, the psychologist William Gaver played environmental sounds to listeners, and asked them to identify what they heard. He found that people are impressive at this: most are capable, for example, of distinguishing running upstairs from running downstairs. Research following in the tradition of work done by the psychologist William H. Warren in the mid-1980s has shown that people are even able to use sound to sense the shapes and textures of some objects.

Our ears and auditory systems are, then, highly designed for and competent at sensing and recognizing what is happening around us. Our auditory systems are priceless pieces of machinery, just the kind of hardware that cultural evolution shouldn't let go to waste, perfect for harnessing. In this chapter, I sift through the sounds of nature and distill a host of regularities found there, regularities that apply nearly anywhere— in the jungle, on the tundra, or in a modern city. The idea is that our auditory system, having evolved in the presence of these regularities for hundreds of millions of years, will have evolutionarily "internalized" them; our auditory system will therefore work best when incoming sounds conform to these regularities. I will then ask whether the sounds of speech across human languages tend to respect these regularities. *That's* what we expect if language harnesses us.

## OVER HEAR

It can be difficult for students to attract my attention when I am lecturing. My occasional glances in their direction aren't likely to notice a static arm raised in the standing-room-only lecture hall, and so they are reduced to jumping and gesturing wildly in the hope of catching my eye. And that's why, whenever possible, I keep the house lights turned off. There are, then, three reasons why my students have trouble visually signaling me: (i) they tend to be behind my head as I write on the chalkboard, (ii) many are occluded by other people, are listening from behind pillars, or are craning their necks out in the hallway, and (iii) they're literally in the dark.

These three reasons are also the first ones that come to mind for why languages everywhere employ audition (with the secondary exceptions of writing and signed languages for the deaf) rather than vision. We cannot see behind us, through occlusions, or in the dark; but we *can hear* behind us, through occlusions, and in the dark. In situations where one or more of these—(i), (ii) and (iii) above—apply, vision fails, but audition is ideal. Between me and the students in my course lectures, all three of these conditions apply, and so vision is all but useless as a route to my attention. In such a scenario a student could develop a firsthand appreciation of the value of speech for orienting a listener. And if it weren't for the fact that I wear headphones blasting Beethoven when I lecture, my students might actually learn this lesson.

The three reasons for vision's failure mentioned above are good reasons why audition might be favored for language communication, but there is a much more fundamental reason, one that would apply to us even if we had eyes in the backs of our heads and lived on wide-open prairies in a magical realm of sunlit nights. To understand this reason, we must investigate what vision and audition are each good at.

Vision excels at answering the questions "What is it?" and "Where is it?", but not "What happened?" Each glance cannot help but inform you about what objects are around you, and where. But nearly everything you see isn't *doing* anything. Mostly you just see nature's set pieces, currently not participating in any event—and yet each one is visually screaming, "I'm here! I'm here!" There's a simple reason for this: light is reflecting off all parts of the scene, whether or not the parts have anything interesting to say. Not only are all parts of a scene sending light toward you even when they are not involved in any event, but the visual stimulus often changes in dramatic ways even when the objects out there are not moving. In particular, this happens whenever *we* move. As we change position, objects in our visual field dynamically shift: their shapes distort, nearer objects move more quickly, and objects shift from visible to occluded and vice versa. Visual movement and change are not, therefore, sure-fire signals that an event has occurred. In sum, vision is not ideal for sensing events because events have trouble visually outshouting all the showy nonevents.

If visual nature is the loquacious coworker you avoid eye contact with, auditory nature is (ironically) the silent fellow who speaks up only to say, "Piano falling." Audition excels at the "What's happening?" sensing a signal only when there's an event. Audition not only captures events we cannot see—like my (fictional) gesticulating students—but serves to alert us to events occurring even within our view. Nonevents may be screaming visually, but they are not actually making any noise, and so audition has unobstructed access to events—for the simple reason that sound waves are cast only when there is an event.

That's why audition, but not vision, is intrinsically about "what's happening." Audition excels at event perception. And this is crucial to why audition, but not vision, is best suited for everyday language communication. Communication *is* a kind of event, and thus is a natural for audition. That is, everyday person-to-person language interactions are acute events intended to be comprehended *at that moment*. Writing is not like this; it is a longer-term record of our thoughts. And when writing *does* try to be an acute person-to-person means of communication, it tends to take measures to ensure that the receiver gets the message *now*—and often this is done via an *auditory* signal, such as when one's e-mail or text messaging beeps an alert that there is a new message.

That language is auditory and not visual is, in the broadest sense, a case of harnessing, or being like nature for the purpose of best utilizing our hardware. Language was culturally selected to utilize the auditory modality because sound is *nature's* modality of event communication.

That's nice as far as it goes, but it does not take us very far. The Morse code for electric telegraphy utilizes sound (dots and dashes), and even the world-record Morse code reader, Ted McElroy, could only handle reading 75.2 Morse code words per minute (a record set in 1939), whereas we can all comprehend speech comfortably at around 150 words per minute—and with effort, at rates approaching 750 words per minute. Fax machines and modems also communicate by sound, but no human language asks us to squeal and bleep like that. Clearly, not just any auditory communication will do. And that brings us to the main aim of this chapter: to say what auditory communication *should* sound like in order to best harness our auditory system. We move next to the first step in this project: searching for the *atoms* of natural sounds, akin, to the contours in natural scenes on the visual side.

## NATURE'S PHONEMES

By understanding the different evolutionary roles for vision and audition, we just saw that audition is the appropriate modality to harness for language: sound is nature's standard event stream, and language therefore wants to utilize sound to make sure language utterances get received. But what kinds of sounds, more specifically, should language use to best harness our brains? The sounds of nature, of course. But the natural world has a large portfolio of sounds it can make, and people are good at mimicking a fair share of these sounds, mostly with their mouths, but sometimes with the help of their hands and underarms. Saying that a well-designed language will use sounds from nature is like saying one had "a sandwich" in a deli. *Which* sounds from nature? Wind blowing, water splashing, trees falling (when someone is around), leaves rustling, thunder, animal vocalizations, knuckle cracks, eggs breaking? Where is language to begin?

Although nature's sounds are all over the map, there's order to the cacophony. Most events we hear are built out of just three fundamental building blocks: hits, slides, and rings.

Hits happen whenever a solid object bumps into another object. When you walk, your feet hit the ground. When you knock, your knuckles hit the door. A tennis match is a game of hits—ball hits racket, ball hits net, ball hits ground. Hits make a distinctive sound. They happen suddenly, and the auditory signal consists of an almost instantaneous explosive burst of energy emanating from the impact.

Slides are the other common kind of physical interaction between solid objects. Slides occur whenever there is a long duration of friction contact between surfaces. If you drag your finger down the page of this book, you're making a slide. If you push a box along the floor, that's a slide. The auditory structure of slides differs from that of hits: rather than a nearly instantaneous release of energy, slides have a non-sudden start and a white-noise-like sound that can last for a more extended period of time. Slides are less common than hits. First, they require a special circumstance, the extended interaction of two surfaces; hits, on the other hand, are what perception scientists call "generic," because no special coincidences are needed to carry off a hit. Second, when slides *do* happen their friction tends to significantly lower the energy in the event, and therefore they commonly occur at the tail ends of events. Third, whereas a long sequence of hits is possible (with intervening rings, as discussed in a moment)—as when a ping pong ball bounces lower and lower, for instance—a long sequence of distinct slides is not typically possible; something would have to stop one slide to allow another one to start, but any such interference with a slide is likely to involve a hit.

Hits and slides are the only physical interactions among solid objects that we regularly experience, and they are certainly the primary ones our ancestors would have experienced. We are land mammals. Splashes, involving a solid and a liquid, are neither hits nor slides, and although they could shape the auditory system of otters, seals, and whales, they're unlikely to be of central significance to *our* auditory system.

With the two kinds of solid-object physical interaction out

of the way, we are left with the final fundamental constituent of these natural events: rings. A ring is what happens to a solid object *after* a physical interaction, that is, after a hit or a slide. When a solid object is physically impinged upon, it vibrates and wobbles, and although one can almost never see these vibrations, one can hear them. You can tell from the sound whether your pen is tapping your desk, your computer, or your coffee mug, because the same pen hit leads to different rings; you may also be able to tell that it is the *same* pen hitting the three different objects.

Different objects ring in distinct "timbres," a word (pronounced "TAM-ber") which refers to the overall perceptual nature of the sound. For example, a piano C and a violin C have the same pitch, or frequency, but they differ in the quality or texture of their sound, and timbre refers to this. Most objects have very short-lived rings—unlike the long-drawn-out ring of a gong—but they do ring, and once you set your mind to noticing, you'll be amazed to hear these rings everywhere. And it is not just hits that ring, but slides as well. The vibrations that occur when any two objects hit each other will have many similarities to the vibrations resulting from the same two objects sliding together, so that we can tell that a coffee mug is being dragged along the desk because the ring possesses certain features also found in the ring of a pinged coffee mug.

Hits, slides, and rings are, therefore, nature's primary phonemes (see Figure 3). They are a consequence of how solid physical objects interact and vibrate. Although these three kinds of sound are special in the lexicon of nature, there is nothing requiring *language* to carve sounds at these joints. Dog woofs, cat calls, horse neighs, whale song, and bird song do not carve at these joints. Neither does the auditory communication of a fax machine. But if a language is to be designed to harness the human auditory system, then it will be built out of the sounds of hits, slides, and rings.

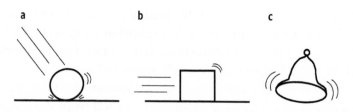

**FIGURE 3.** The three principal constituents of physical events: **(a)** hits, **(b)** slides and **(c)** rings. They sound suspiciously similar to plosives, fricatives and sonorant phonemes in human languages.

Are human languages built out of these constituents? Yes. In fact, the most fundamental universal of human speech is that phonemes, the "atoms" of speech, come in three primary types, and these types match nature's phonemes! Language's hits, slides and rings are, respectively, plosives, fricatives, and sonorants.

Plosives—like *b, p, d, t, g,* and *k*—are found in every language, and consist of sudden, explosive, high-energy inceptions. Plosives sound like hits (even embedding their ex*plosive* hit-like starts in the name). Figure 3a, on the left, shows the time-varying frequency distribution for the sound made when I hit my desk with a small plastic cup, and one can see that the hit begins with a sharp vertical line indicating the presence of a wide range of frequencies at the instant of the collision. That same figure shows, on the right, the same kind of plot when I made a "k" sound. Again one can see the sharp edge at the beginning of the sound, characteristic of a hit. (Also note that, in English, at least, one finds many plosive-filled words with meanings related to hits: bam, bang, bash, blam, bop, bonk, bump, clack, clang, clink, clap, clatter, click, crack, crush, hit, klunk, knock, pat, plunk, pop, pound, pow, punch, push, rap, rattle, tap, and thump.)

Languages have a second principal kind of consonant called the fricative, such as *s, sh, th, f, v,* and *z.* They are extended and noisy, and sound like slides. (In fact, the very word "fricative" captures the friction nature of a slide.) And just as slides are rarer than hits, fricatives are less common than plosives. All languages have plosives, whereas many languages (especially in Australia) do not have fricatives. Figure 4b, on the left, shows the frequencies of sound emanating from a small cup that I slid on my desk, and one can see that there is no longer a crisp start to the sound as there was for hits. There is also a longer duration of sound, all of it with a wide range of frequencies. On the right of Figure 4b is the same kind of plot, this one generated when I made a "sh" sound. One sees the signature features of a slide in fricatives. (Also note that in English, at least, one finds many fricative-filled words with meanings related to slides: fizzle, hiss, rustle, scratch, scrunch, shuffle, sizzle, slash, slice, slip, swoosh, whiff, whiffle, and zip.)

The third principal phoneme type used across human languages is the sonorant, including vowels like *a, e, i, o, u,* but also sonorant consonants like *l, r, y, w, m,* and *n.* Each of these phonemes has strongly periodic vibrations, and has a complex spectral shape. Sonorants sound like rings. Figure 4c, left, shows the ringing after tapping my coffee mug. Only certain frequencies occur during the quickly decaying ring, and these frequency bands are characteristic of the shape and material properties of my mug. To the right of that in Figure 4c is the signal of me saying "ka." (The plosive "k" sound corresponds to the tap.) As with the coffee mug, there are certain frequency bands that are more active, and these patterns are what characterize the sound as an "a."

Lo and behold! The principal three classes of phonemes in human speech sound just like nature's three classes of phoneme. We speak in hits, slides, and rings!

Before getting overly excited by the realization that language's phonemes are like nature's phonemes, we must, however, address a worry: How else *could* we speak? What if human vocalization can't *help* but sound like hits, slides, and rings? If that were the case, then the observations made in this section would have little significance for harnessing; culture would not need to design language to sound like hits, slides, and rings, because our mouths would make these sounds by default. We take this up next.

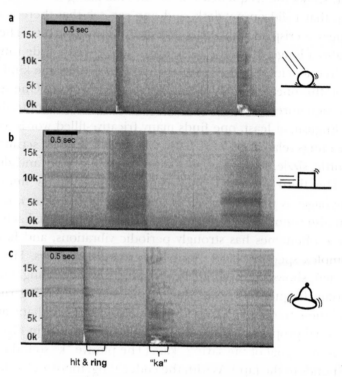

**FIGURE 4**. Illustration that plosives, fricatives and sonorants sound like hits, slides and rings, respectively. These plots show the frequencies on the y-axis, and time on the x-axis. Comparison of **(a)** hits and plosives, **(b)** slides and fricatives, and **(c)** rings and sonorants.

## TONGUE WAGGING

When the Mars Rover landed on Mars, it bounced several times on balloon-like cushions; the cushions then deflated, allowing the rover to roll gently onto the iron-red dirt. If you had been there watching the bouncy landing, you would have heard—as you writhed in pain from decompression in the low-pressure atmosphere—a sequence of hits, with rings in between. And once the rover found a place to take a sample of Martian soil, it would have scraped debris into a container for analysis, and that scrape would have sounded like a slide, followed by a ring characteristic of the Rover's scraping arm. Hits, slides, and rings on Mars! It is not so much that hits, slides, and rings are Earthly *nature's* phonemes as that they are *physics's* phonemes. These sounds are the principal building blocks of event sounds anywhere there are solid objects interacting—even in our mouths.

Our mouths have moving parts, including a powerful and acrobatic tongue; fleshy, maneuverable lips; and a jaw rigged with rock-hard teeth. When we speak, these parts physically interact in complex ways, creating speech events. But speech events are *events*, and if hits, slides, and rings are the fundamental constituents of physical events, then speech events must *also* be built from hits, slides and rings in the mouth. It is no wonder, then, that human speech sounds like hits, slides and rings. Speech is built from the fundamental constituents of physical events because speech *is* a physical event. Harnessing would appear to have nothing to do with it.

However, when we speak, our mouth is not simply a container with a tongue, lips, and teeth rattling around. We are not, for example, making hit sounds by tapping our teeth together, or slide sounds by grinding our teeth. When our mouth (in collaboration with our nose, throat, and lungs) makes sounds, it is using mechanisms for sound production that go well beyond the solid-object event atoms—hits, slides, and rings. Although hits, slides, and rings are the most fundamental kinds of physical events (because solid-object events are the most fundamental kind of physical event), they are not the *only* kinds. There are hosts of others. In particular, there are many physical events that involve the flow of fluid or air. The events in our mouths that make the sounds of speech are events involving airflow, not hits, slides, or rings at all. Airflow events in our mouths *mimic* hits, slides, and rings, the constituents of solid-object physical events. Our mouths make a plosive by a sudden release of air, not by an actual collision in the mouth! Fricatives are made by the non-instantaneous movement of air through a tight passage; no surfaces in the mouth are actually rubbed against one another. And sonorants are not due to an object vibrating because of a hit or slide; instead, sonorants come from the vocal chords vibrating as air passes by.

Hit, slide, and ring sounds without hits, slides, or rings! What a coincidence! Human speech employs three principal sounds via airflow mechanisms, and yet they happen to sound just like the three principal sounds that happen in events with physical interactions between solid objects. Utterly different mechanisms, but the same resultant sound. That's too coincidental to *be* a coincidence. That's just what harnessing expects: airflow sound-producing mouths settling on just a few sounds for language—the sounds of physical interactions among solid objects.

We must be careful, though. What if airflow mechanisms cannot help but make hit, slide, and ring sounds? Or, more to the point, could it be that the particular airflow mechanisms our mouths are capable of can lead *only* to sounds like hits, slides and rings? No. Human mouths are capable of sounds much more varied than the sounds of interacting solid objects. For example, people can mimic many animal sounds—quacks, moos, barks, ribbits, meows, and even human sounds like slurps, burps, sneezes, and yawns—that are constructed out of constituents beyond simple hit, slide and ring sounds. People can mimic water-related sounds— like splashes, flushes, and drips—none which are built from hit, slide, and ring sounds. And our airflow sound-mimicking mouths can, of course, mimic airflow sounds—like a soda

pop being opened, howling wind, or even breaking wind—
also unrelated to the sounds of hits, slides, and rings. People
can mimic "hot" sounds, like sizzling bacon and roaring fires.
They can even mimic the sounds of revving motorcycles, fax
machines, digital alarm clocks, shrilling phones, and alien
spaceships, none of which are sounds built from hits, slides,
and rings. We see, then, that our airflow sound-producing
mouths have a very wide repertoire, and yet speech has
employed only the barest of our talents for mimicry, prefer-
ring exactly the sounds that occur among interacting macroscopic solid objects. We're not, therefore, speaking in hits,
slides, and rings by default. That we find these in all languages
is a sign that we have been harnessed.

In upcoming sections, I will also concentrate on some other
kinds of sounds our mouths can produce, but that language
tends to avoid; these cases deserve special attention because
of their prima facie similarity to sounds we *do* find in speech.
Thus, they can help to answer the question of why speech utilizes some sounds we can make, but not others we can make
just as easily. For example, we will see in the upcoming section
that although we can make the sounds of *wiggly* hits and slides,
we do not have them as phonemes—and this is consistent with
their absence in physics. In the section following that we will
see that although we can make slide-hit sounds and hit-slide
sounds, only the latter is given the honor of phoneme status in
languages (see the section titled "Nature's Other Phoneme"),
consistent with hit-slides being a fundamental sound in
physics, while slide-hits are not. And we'll see in the "Two-Hit
Wonder" section that a simple kind of sound (a "beep") that
could exist as a phoneme does not occur in human languages,
consistent with its non-primitive status in physics. More generally, for the next five sections I will brandish a magnifying glass
and closely examine the internal structures of hits, slides, and
rings, asking whether those same fine structures are found in
plosives, fricatives, and sonorants, respectively.

## WIGGLY RINGS

Harmonicas don't get no respect. They're cheap (I just found one online for $5), tiny hunks of metal that tend to be played by guys who didn't finish finishing school. I've had a couple of harmonicas for years, and have never understood them: they don't have all the notes and can only play three chords. Blowing on a harmonica can't help but sound fairly good, but I have always been frustrated by my inability to get it to do much more. A serious blues harmonica player can create sounds far richer than seems possible from what would appear to be little more than a toy.

A harmonica is deceptive because it is, in a sense, not an entire instrument at all. It is perhaps half an instrument—maybe that's why they're so inexpensive. The other half of the instrument is the human hand. That explains why the best harmonica players have hands, and, in addition, tend to move them all about the instrument when playing. This is described as "bending" the notes, and by doing so, the performer can provide a musical dynamism not possible with just the twenty or so notes in the harmonica's range. The sounds reaching the listener's ears are not only those coming directly from the harmonica, but also the harmonica sounds that first bounce off objects in the environment before reflecting toward the listener's ears. For the note-bending blues performer, the hands are the objects the sounds bounce off. Each time a sound bounces off something, some sound frequencies are absorbed more than others, and so the timbre of the sound coming from that reflection is changed. The total timbre depends on the totality of harmonica sounds that reach the ear directly and indirectly from all points in the environment. And we're able to hear these sound shapes, which is why harmonica benders go to all the trouble of wiggling their hands—and why there are acoustics engineers who worry about the physical layout of auditoriums.

Bending and acoustic reflections don't just matter in the blues and in concert halls where instruments (including half instruments) are crooning out musical tones. Objects involved in events also croon, or ring. A ring has a complex timbre that informs us of the object's size, shape, and material. But just like harmonica sounds, rings can get bent by the environmental surroundings. And our brains can decode the bends, and can give us a sense of our surroundings purely on the basis of the shapes of the sounds reaching our ears. The psychologist James J. Jenkins demonstrated in 1985 that blindfolded students, after a little practice, can navigate very well amongst obstacles by utilizing such auditory cues.

These acoustical observations about how the surroundings affect sound have an important consequence for the internal structure of rings: rings can be wiggly. There are several converging reasons for this. First, an event that causes a ring often also sets the ringing object in motion: something has been hit, or something is sliding. Because the shape of a ring reaching one's ears depends on the object's surroundings, ringing objects that are moving produce rings that vary over time. Second, when an event occurs, we are often on the move. Because the shape of the ring we receive depends, in part, upon our position in the world, the shape of the ring reaching our ears may be varying over time. In each case, whether we are moving or the object is, the timbre of a ringing object can change, and these are wiggles we notice, at least subconsciously. In addition to such dynamic changes in the subtleties of a ring's timbre, there is another dimension in which rings can often vary: pitch, the musical-note-like, "higher" or "lower" quality of sound. When motion is involved—either our own motion or that of the objects involved in events—we get Doppler shifts, a phenomenon we are all familiar with, as when a car approaching you sounds higher-pitched than when it is moving away. (See also the later section of this chapter titled "Unresolved Questions" for more about the Doppler effect and its stamp upon speech. And see the following chapters on music, where the Doppler effect will be discussed in detail.)

Rings can therefore change over time, both in timbre and in pitch. That is, a single ring can often be intrinsically dynamic. What about hits and slides?

Hits are nearly instantaneous, and for this simple reason they cannot change over time, at least not in the sense of continuously varying from one kind of hit to another. Hits can, of course, happen in quick succession, such as when you drop a pen and one end hits an instant before the other. But such a pen event would be *two* physical interactions, not one. Unlike a single ring, which can wiggle, a single hit has no wiggle room.

How about slides? Slides can occur for a lot longer than an instant, and so they can, in principle, dynamically vary over their occurrence. Although slides can be long—for example, a single snowy hill run on a sled may be one continuous slide— they are much more commonly short (though not instantaneous) in duration, because they quickly dissipate the energy of an event, sometimes ending it. Do the sounds of slides ever, in fact, dynamically vary over time? Before answering this, let's be clear on what we mean by the sound of a slide. A slide can cause a ring, as we have discussed, but that is not what we're interested in at the moment. We are, instead, interested in the sound made by the physical interaction of the two sliding surfaces—the noisy friction sound itself, caused by the coarseness of the objects involved. Therefore, to produce a wiggly slide, the coarseness of the surface being slid upon would have to vary, so that one friction sound would change gradually to another friction sound. Although coarseness varies randomly on lots of materials, few objects vary in a systematic, graded fashion, and thus slides will tend to have a rather non-varying sound.

Rings, then, can be wiggly. But not hits, and not slides. If language has culturally evolved to sound like nature, then we would expect that sonorant phonemes (language's rings) would sometimes be dynamically varying, but not plosives (language's hits) or fricatives (language's slides).

Languages do, indeed, often have sonorants that vary during their utterance. Although vowels like those in "sit" and in "set" are non-varying, some vowels do vary, like those in "skate" and "dive." When one says "skate," for example, notice how the vowel sound requires your mouth to vary its shape, thereby dynamically modulating its timbre (in particular, modulating something called the *formant structure*, where formants are the bands of frequencies emanating from a sonorant). Vowel sounds like these are called diphthongs. Furthermore, sonorant consonants like *l*, *r*, *y*, *w*, and *m* demand ring changes. For example, when you say "yet," notice how during the "y" your mouth dynamically varies its shape. These sonorants incorporate timbre changes. Recall that rings in nature also can change in pitch due to the Doppler effect. Do we find something like the Doppler shift in sonorant phonemes? Yes, in fact, in the many tonal languages of the world (such as Chinese), where vowels may be distinguished from one another only by virtue of how they dynamically vary their pitch during their utterance.

Whereas sonorants are commonly wiggly, effectively making more than one ringing sound during their utterance, no language possesses phonemes having in them more than one hit sound. It is possible in principle to have a single phoneme that sounds like two hits in very quick succession—for example, the "ct" in "ectoplasm"—but while we can make such sounds, and they even occur in language, they are never given building-block, or phoneme, status.

Are language's slides like nature's slides in being non-wiggly? First, let's be clear on what it would even mean to have a fricative that varies dynamically as it is spoken. Try saying the sound "fs." That is, begin with an "f" sound, and then slowly morph it to become "s" at the end. You make this sound when, for example, you say "puffs." Languages could, in principle, have fricative phonemes that sound like "fs." That is, languages could possess a *single* phoneme that has this complex dynamic fricative sound, just as languages possess single sonorant phonemes that are dynamic. One does not, however, find phonemes like this among human languages.

Nature's rings are wiggly but hits and slides are not, and culture has given us language with the same wiggles: language commonly has sonorant phonemes that dynamically vary, but does not have plosive or fricative phonemes that dynamically vary. Our auditory systems are happy with dynamic rings, but not with dynamic hits or slides, and culture has given us speech that conforms to these tastes.

In addition to looking at dynamic changes within phonemes, we can make similar observations at the level of how phonemes combine into words: languages commonly have words with multiple sonorants in a row, but more rarely have multiple plosives or multiple fricatives in a row. For example, consider the following English words, which I found by perusing the first paragraph of this chapter: "harrowing" possesses six sonorants in a row (a, rr, o, w, i, and ng, the latter which is a nasal sonorant), "village" has three in a row, "generation" has five in a row, and "eventually" has four in a row. One *can* find adjacent plosives, like in "packed" ("kt") and "grabbed" ("bd"), and one can find adjacent fricatives like in "puffs" ("fs"), "gives" ("vz") and "isthmus" ("sth"), but finding more than two in a row is difficult, and five or six in a row is practically impossible.

We now know how, and how much, each of the three kinds of "event atoms" can vary in sound while they are occurring. We have not, however, considered whether an event of one of these three kinds can ever dynamically change into another *kind* of event. Could some simple event pairs be so common that we are likely to possess special auditory mechanisms for their recognition, mechanisms language harnesses? We turn to this question next, and uncover a kind of event sufficiently fundamental in physics that it is also found as a fourth kind of phoneme in language.

## NATURE'S OTHER PHONEME

I have been treating hits and slides as two different kinds of physical interaction. But slides are more complex than hits. This is because slides consist of very large numbers of very low-energy hits. For example, if you rub your fingernail on this piece of paper, it will be making countless tiny collisions at the microscopic level. Or, if you close this book and run your fingernail over the edges of the pages of the book, the result will be a slide with one little hit for each page of the book. But it would not be sensible to conclude, on this basis, that there are just two fundamental natural building blocks for events—hits and rings—because describing a slide in terms of hits could require a million hits! We still want to recognize slides as one of nature's phonemes, because slides are a kind of super-sequence of little hits that is qualitatively unlike the hits produced when objects simply collide.

But there are implications to the fact that slides are built from very many hits, but not vice versa: that fact opens up the possibility of a fundamental event type that is not quite a hit, and not quite a slide. To understand this new event type, let's look at a slide at the level of its million underlying hits. Imagine that the first of these million hits is appreciably more energetic than the others. If this were the case, then the start of the slide would acquire a crispness normally found in hits. But this hit would be just the first of a long sequence of hits, and would thus be part of the slide itself. Such a hit-slide would, if it existed, be neither a hit nor a slide.

And they *do* exist, for several converging reasons. First, slides have a tendency to be initiated by hits. Try sliding this book on a desk. The first time you tried, you may have bumped your hand into the book in the process of attempting to make it slide. That is, you may have hit the book prior to the slide (see Figure 6). It requires careful attention to gently touch the book without hitting it first. Now grab hold of the book and try to slide it *without* an initial hit. Even in this case there can often be an initial hit-like event. This is because in order to slide an object, you must overcome static friction, the "sticky" friction preventing the initiation of a slide. This initial push is hit-like because the sudden overcoming of static friction creates a sudden burst of many frequencies, as in Figure 2a. Slides, then, often begin with a hit. Second, hits often have slides following them. If you hit a wall with a straight jab, you will get a lone hit, with no follow-up slide. But if you move your arm horizontally next to the wall as you are hitting it—in order to give it a more glancing blow—there will sometimes be a small skid, or slide, after the initial hit.

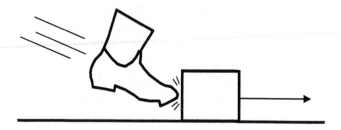

**FIGURE 5.** A hit-slide is a fourth fundamental constituent of physical events. It sounds like a kind of phoneme in language called the affricate, which is like a plosive followed by a fricative.

Although a hit followed by a slide is a natural regularity in the world, a slide followed by a hit is *not* a natural physical regularity. First, it is common to have a hit *not* preceded by a slide. To see this, just hit something. Odds are you managed to make a hit without a slide first. Second, when there is a slide, there is no physical regularity tending to lead to a hit. Slides followed by hits are possible, of course—in shuffleboard, for example (and note the fricatives in "shuffle")—but they really are two separate events in succession. A hit-slide, on the other hand, can effectively be a *single* event, as we discussed a moment ago.

If language sounds like nature, then we should expect linguistic hit-slide sounds to be more common than slide-hit sounds. Later in this chapter—in the section titled "Nature's Words" —I will provide evidence that this is true of the way phonemes combine into words across human languages. But in this section I want to focus on the single-phoneme level. The question is, since hit-slides are a special kind of fundamental event atom, but slide-hits are not, do we find that languages have phonemes that sound like hit-slides, but not phonemes that sound like slide-hits?

Languages, like nature, *are* asymmetrical in this way. There is a kind of phoneme found in many languages called an *affricate*, which is a fricative that begins as a plosive. One example in English is "ch", which is a single phoneme that possesses a "t" sound followed by a "sh" sound. In addition to words like "chair," it also occurs in words like "congratulate" (spoken like "congratchulate"), and often in words like "trash" (spoken like "chrash"). Another example is "j", which begins with "d" sound followed by a voiced version of the "sh" phoneme. Although we can describe "ch" as a "sh" initiated by a "t", it is not the same sound that occurs when we say "t" and quickly follow it up with "sh". The "ch" phoneme has the "t" and "sh" sounds bound up so closely to one another that they sound like a single atomic event. The "tsh" sound in "hotshot," on the other hand, will typically sound different from "ch"; that is, we do not pronounce the word "ha-chot."

Whereas language has incorporated nature's hit-slide phoneme as one of its phoneme types, slide-hits, on the other hand, are *not* one of nature's phonemes, and a harnessing language is *not* expected to have phonemes that sound like slide-hits. Indeed, that is the case. Languages do not have the symmetric counterpart to affricates—phonemes that sound like a plosive initiated by a fricative. It is not that we can't make such sounds. "st" is a standard sound combination in English of this slide-hit form, but it is not a single phoneme. Other cases would be the sounds "fk" and "shp", which occur as *pairs* of phonemes in words in some languages, but not as phonemes themselves.

By examining physics in greater detail, in this section we have realized that there is a fourth fundamental building block of events: hit-slides. And just as languages have honored the other three fundamental event atoms as their principal phoneme types, this fourth natural event atom is also so honored. Furthermore, the symmetrical fifth case, slide-hits, is not a fundamental event type in nature, and we thus expect—if harnessing has occurred—not to find fricative-plosives as language phonemes. And indeed, we don't find them.

## SLIDES THAT SING

Recall that slides are, in essence, built from very many little hits in quick succession. The pattern of hits occurring inside a slide depends on the nature of the materials sliding together, and this pattern is what determines the nature of the slide's sound. If you scrape your pencil on paper, then because the paper's microscopic structure is fairly random, the sound resulting from the many little hits is a bit "noisy," or like radio static, in having no particular tone to it. (The pencil scraping may also cause some ringing in the table or the pencil, but at the moment I want you to concentrate only on the sound emanating from the slide itself.)

However, now unzip your pants. You just made another slide. Unlike a pencil on paper, however, the zipper's regularly spaced ribs create a slide sound that has a tonality to it. And the faster you unzip it, the higher the pitch of the zip. Slides can sing. That is, slides can have a ring-like quality to them, due not to the periodic vibrations of the objects, but to the periodicity in the many tiny hits that make up a slide.

Whether or not a slide sings depends on the nature of the materials involved, and that's why the voice of a slide is an auditory feature that brains have evolved to take notice of: our brains treat singing and hissing slides as fundamentally different because these differences in slide sounds are informative as to the identity of the objects involved in the slides. Although slides can sing, it is more common that they don't, because texture with periodicity capable of a ring-like sound is rare, compared to random texture that leads to generic friction sounds akin to white noise.

Do human languages treat singing slide sounds as different from otherwise similar non-singing slide sounds? Yes. Languages have fricatives of both the singing and the hissing kinds, called the voiced and unvoiced fricatives, respectively. Voiced fricatives include "z", "v", "th" as in "the," and the sound after the beginning of "j" (which you will recall is an affricate, discussed earlier in "Nature's Other Phoneme"). Unvoiced fricatives include "s", "f", "th" as in "thick," and "sh." Just as singing slides will be rarer than non-singing slides— because the former require special circumstances, namely, slides built out of many periodically repeating hits—voiced fricatives are rarer in languages than unvoiced fricatives. John L. Locke tabulated data in his excellent 1983 book *Phonological Acquisition and Change,* and discovered that "s" is found in 172 of 197 languages in the Stanford Handbook* (87 percent) and in 102 of 317 languages in the UCLA Phonological Segment Inventory Database (32 percent), whereas "z" (the voiced version of "s") is found in 77 of 197 languages (39 percent) and 36 of 317 languages (11 percent), respectively. Similarly, "f"

---

* *Handbook of phonological data from a sample of the world's languages: A report of the Stanford Phonology Archive* (1979). Stanford University, Department of Linguistics.

is found in 106 of 197 languages (54 percent) and in 135 of 317 languages (43 percent), whereas "v" is found in 61 of 197 languages (31 percent) and in 67 of 317 languages (21 percent), respectively. These data suggest that unvoiced slides are about twice as likely as voiced slides to be found in a language. (And notice how, in English at least, one finds voiced-fricative words with meanings related to slides that sing: rev, vroom, buzz, zoom, and fizz. One also finds unvoiced-fricative words with meanings related to unsung slides: slash, slice, and hiss.)

Voiced and unvoiced fricatives are found in languages because they're found in the physics of slides. Hits can also be voiced or unvoiced, but for completely different physical reasons than slides. Zip up your pants and let's get to this.

## TWO-HIT WONDER

Each day, more than a billion people wake to the sound of a ringing alarm, reach over, and hit the alarm clock, thereby terminating the ring and giving themselves another five minutes of sleep. In these billion cases a hit *stops* a ring, rather than starting one as we talked about earlier. Of course, the hit on the clock *does* cause periodic vibrations of the clock (and of the sleeper's hand), but the sound of these vibrations is likely drowned out by the sound of the alarm still ringing in one's ears.

Although hitting the snooze button of an alarm clock is not a genuine case of a hit stopping a ring, there *are* such genuine cases. Imagine a large bell that has been struck and is ringing. If you *now* suddenly place your hand on it, and keep it there, the ringing will suddenly stop. Such a sudden hand placement amounts to a hit—a hit that sticks its landing. And it is, in this case, by virtue of dampening, a hit that leads to the *termination* of a ring. Some dampening will occur even if your hand doesn't stick the landing, so long as you hit the bell much less energetically than it is currently ringing; the temporary contact will "smother" some of the periodic vibrations occurring in the bell.

Although in such cases it can sound as if the bell's ringing has terminated, in reality one can leave the bell with a residual ring. A hit on a quiet bell would sound explosive hit, because in contrast to the bell's stillness, the hit is a sudden discontinuous *rise* in the ringing magnitude. But that same hit on an already very loudly ringing bell causes a sudden discontinuous *drop* in the ringing magnitude. In contrast to the loud ringing before the hit, the hit will sound like the sudden *ceasing* of a ring, even if there is residual ringing.

Hits, therefore, have two voices, not just the one we discussed earlier in the section called "Nature's Phonemes." Hits not only can create the sudden appearance of a wide range of frequencies, but can *also* sometimes quite suddenly dampen out a wide range of frequencies. These two sounds of hits are, in a sense, opposites, and yet both are possible consequences of one and the same kind of hit. This second voice is rarer, however, because it depends on there already being a higher-energy ring before the hit, which is uncommon because rings typically decay quickly. That is, the explosive voice of hits is more common than the dampening voice, because most objects are not already ringing when they are hit.

If languages have harnessed our brain's competencies for natural events, then we might expect languages to utilize both of these hit sounds. And indeed they do. The plosives we discussed earlier consisted of an explosive release of air, after having momentarily stopped the airflow and let pressure build. But plosives also occur when the air is momentarily stopped, but not released. This happens most commonly when plosives are at the ends of words. For example, when you utter "what" in the sentence, "What book is this?", your mouth goes to the anatomical position for a "t," but does not ever release the "t" (unless, say, you are angry and slowly enunciating the sentence). Such instances of plosive stop sounds are quite common in language, but less so than released plosive sounds—there are many languages that do not allow unreleased plosives, but none that do not allow released plosives. John Locke tabulated from the Stanford Handbook that, in 32 languages that possessed word-position information (i.e., where plosives can occur within words), no plosives were off limits at word starts (where they would be released), but 79 plosives were impermissible at word-final position (where they are typically unreleased). Also, among the words we collected from 18 languages, 16,130 of a total of 18,927 plosives, or 85 percent, were directly followed by a sonorant (and thus were released), and therefore only 2,797 plosives, or 15 percent, were unreleased. And even in languages (like English) that allow both kinds of plosive sounds, plosives are more commonly employed in their explosive form, something we will talk about in a later section ("In the Beginning"). This fits with the pattern in nature, where explosive hits are more common than dampening hits.

Not only does language have both hit sounds as part of its repertoire, but, like nature, it treats the unreleased "t" sound and the released "t" sound as the *same* phoneme. This is remarkable, because they are temporal opposites: one is like a little explosion, the other like a little *anti*-explosion. One can imagine, as a thought experiment, that people could have ended up with a language that treats these two distinct "t" sounds as two distinct phonemes, rather than two instances of a single one. In light of the auditory structure of nature, however, it is not at all mysterious: any given hit can have two very different sounds, and language carves at nature's joints.

In light of the two sounds hits make, there is a simple kind of sound we can make, but that language never includes as a phoneme: "beep," like an electronic beep or like Road Runner. A beep consists of a sudden start of a tone, and then a sudden stop. Beeps might, at first glance, seem to be a candidate for a fundamental constituent of communicating by sound: what could be simpler, or more "raw", than a beep? However, although our first intuitions tell us that beeps are simple, in physics they are not. In the real world of physical events among objects, beeps can only happen when there is a hit (the abrupt start to the beep), a ring which follows (the beep's tone), and a second hit, this one a dampening one (the abrupt beep ending). A "simple" beep can't happen in everyday physics unless *three* simple constituent events occur. And we find that in languages as well: there are no beep-like phonemes. To make a beep sound in language requires one to first say a plosive of the released kind, then a (non-wiggly) sonorant, and finally an unreleased plosive . . . just like when we say the word "beep."

## HESITANT HITS

Bouncing a basketball could hardly be a simpler event. A bounce is just a hit, followed by a ring. And as we discussed earlier, the sound is a sudden explosion of many frequencies at the initiation of the hit, followed by a more tonal sound with a timbre due to the periodic vibrations of the basketball and floor. Although hits seem simple, they become complicated when viewed in super slow motion. After the ball first touches the ground, the ball begins to compress, a bit like a spring. After compression, the ball then decompresses as it rises on its upward bounce. Although these ball compressions and decompressions are typically very fast, they are not instantaneous: the physical changes that occur during a hit occur over an extended period of time, albeit short. What happens during this short period of time depends on the nature of the objects involved.

One of the most important acoustical observations about collisions is that ringing doesn't tend to occur until the collision is entirely finished. There are several reasons why this

is so. First, the ground rings less during the collision because
even though the ground has already been struck, the ball's
contact with the ground dampens the ground's vibrations.
Similarly, the ground's contact with the ball dampens the *ball's*
vibrations. Second, during the ball's compression, its shape
is continually varying, and so any vibrations it is undergoing
are changing in their timbre and pitch very quickly, far more
quickly than the ring-wiggles we discussed earlier. In fact, the
vibration changes occur at a time scale so short that any rings
that do occur during the collision will not sound like rings at
all. Third, during the period of the collision when the ball is
not yet at maximum compression, the ball is continually hitting new parts of the ground. This is because, as the ball com-
presses, the ball's footprint on the ground keeps enlarging,
which means that new parts of the ball continually come into
contact with the ground. In fact, even if the surface area of
contact never enlarges, the mass in parts of the ball continues
to descend during the ball's compression, providing further
impetus upon the surface area of contact. Because the compression period is filled with many little hits, any ringing occurring during compression will have a tendency to be drowned
out by the little hits.

For several converging reasons, then, the ringing that
occurs after a hit doesn't tend to begin until the compressions
and decompressions are over. For the basketball, the ringing
occurs most vigorously when the ball rebounds back into the
air. There is a simple lesson from these super slow motion
observations: *there is often a gap between the time of the start of a
collision and the start of the ringing.*

What determines the length of these hit-to-ring gaps? When
your basketball is blown up fully, and the ground is firm, then
the time duration of the contact with the floor is very short,
and so the gap between the start of the hit and the ring is very
small. However, when the ball is fairly flat—low in pressure—it
spends more time interacting with the ground. Bouncing a

ball on soft dirt would also lead to more ground time (see Figure 6). Figure 7a shows the sound waveform signal of a book falling onto a crumpled piece of paper—producing similar acoustics (in the relevant respects here) to those of a ball dropped on soft dirt—and one can see a hit-to-ring gap that is larger than that for the same book falling directly onto the table. For the flat basketball, then, the gap between hit and ring is larger than that for the properly blown-up ball.

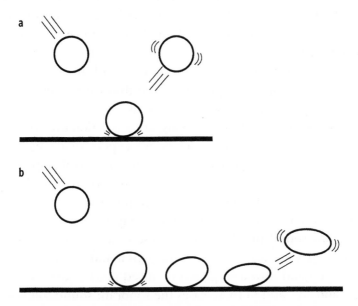

**FIGURE 6. (a)** A rigid hit (i.e., involving rigid objects) rebounds – and rings – with little delay after the initial collision. **(b)** A non-rigid hit takes some time before rebounding and ringing. These physical distinctions are similar to the voiced and unvoiced plosives.

The key difference between the high-pressure ball and the flat ball—and the difference between the book falling on a solid desk versus crumpled paper—is that the former is more rigid than the latter. The more rigid the objects in a collision, the shorter the compression period, and the shorter the gap between the initial hit and the ring. The high-pressure ball is not only more rigid than the flat ball, but also more elastic. More-elastic objects regain their original shape and kinetic energy after decompression, lose less energy to heat during compression, and tend to have shorter gaps. Also, if an object breaks, cracks, or fractures as it hits—a kind of nonrigidity and inelasticity—the gap is longer.

Therefore, although some hits ring with effectively no delay, other kinds of hits take their time before ringing. Hits can be hesitant, and the delay between hit and ring is highly informative because it tells us about the rigidities of the objects involved. Our auditory systems understand this information very well: they have been designed by evolution to possess mechanisms for sensing this gap, and thus for perceiving the rigidity of the objects involved in events.

Because our auditory systems are evolutionarily primed to notice these hit-to-ring delays, we expect that languages should have come to harness this capability, so that plosives may be distinguished on the basis of such hit-to-ring delays. That is, we would expect that plosive phonemes will have as part of their identity a characteristic gap between the initial explosive sound and the subsequent sonorant. Language does, indeed, pay homage to the hit-ring gaps in nature, in the form of voiced and unvoiced plosives. Voiced plosives are like "b," "g," and "d," and in these cases the sonorant sound following them occurs with negligible delay (Figure 7b, left). They even *sound* bouncy—"boing," "bob," and "bounce"—like a properly inflated basketball. Unvoiced plosives are like "p," "k," and "t," and in these cases there is a significant delay after the plosive and before the sonorant sound begins, a delay called the *voice onset time* (Figure 7b, right). (Try saying "pa," and listen for when your voice kicks in.) In English we have short voice onset times and long ones, corresponding to voiced and unvoiced plosives, respectively. Some other languages have plosives with voice onset times in between those found in English.

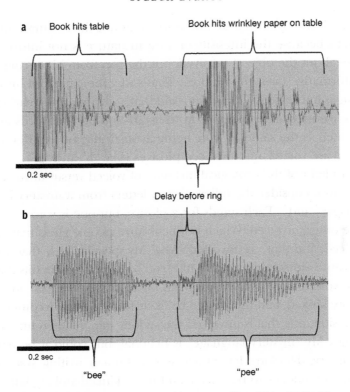

**FIGURE 7.** Illustration that voiced plosives are like rigid, elastic hits, and unvoiced plosives like nonrigid, inelastic hits. These plots show the amplitude of the sound on the y-axis, and time on the x-axis. **(a)** The sound made by a stiff hard cover book landing on my wooden desk on the left, followed by the sound of that same book landing on my desk, but where a wrinkley piece of paper cushioned the landing (making it less rigid and less elastic). **(b)** Me saying "bee" and "pee". Notice that in the inelastic book-drop and the unvoiced plosive cases – i.e., the right in **(a)** and **(b)** – there is a delay after the initial collision before the ringing begins.

Not only do languages utilize a wide variety of voice onset times—hit-to-ring gaps—for plosive phonemes, but one does not find plosive phonemes that don't care about the length of the gap. One could imagine that, just as the intensity of a spoken plosive doesn't change the identity of the plosive, the voice onset time after a plosive might not matter to the identity of a plosive. But what we find is that it always does matter. And that's because the intensity of a hit in nature is not informative about the objects involved, but the gap from hit to ring *is* informative (as is the timbre). That's why the gap from hit to ring is harnessed in language. And that's why, as we saw earlier, the distinct plosive sounds at the start and end of words are treated as the same, despite being acoustically more different than are voiced and unvoiced plosives (like "b" and "p").

In light of the ecological meaning of voiced versus unvoiced plosives, consider the following two letters from a mystery language:   and  . Each stands for a plosive, but one is voiced and the other unvoiced. Which is which? Most people guess that   is voiced, and that   is unvoiced. Why? My speculation is that it is because   looks rigid, and would tend to be involved in hits that are voiced (i.e., a short gap from hit to ring), whereas   looks more kinked, and thus would be likely to have a more complex collision, one that is unvoiced (i.e., a long gap between hit and ring). My "mystery language" is fictional, but could it be that more rigid-looking letters across real human writing systems have a tendency to be voiced, and more kinked-looking letters have a tendency to be unvoiced? It is typically assumed that the shapes of letters are completely arbitrary, and have no connection to the sounds of speech they stand for, but could it be that there are connections because objects with certain shapes tend to make certain sounds? This is the question Kyle McDonald—a graduate student at Rensselaer Polytechnic Institute (RPI) working with me—raised and set out to investigate. He found that letters having junctions with more contours emanating from them—i.e., the more kinked letters—have a greater probability of being unvoiced. For example, in English the three voiced plosives are "b," "d," and "g," and their unvoiced counterparts are "p," "t," and "k." Notice how the unvoiced letters—the t and k, in particular—have more complex structures than the voiced ones. Kyle McDonald's data—currently unpublished—show that this is a weak but significant tendency across writing systems generally.

## RIGID MUFFLER

As I walk along my upstairs hallway, I accidentally bump the hammer I'm carrying into the antique gong we have, for some inexplicable reason, hung outside the bedroom of our sleeping infant. I need to muffle it, quickly! I have one bare hand, and the other wielding the guilty hammer; what do I do? It's obvious. I should use my bare hand, not the hammer, to muffle the gong. Whereas my hand will dampen out the gong ring quickly, the hammer couldn't be worse as a dampener. My hand serves as a good gong-muffler because it is fleshy and nonrigid. My hand muffles the gong faster than the rigid hammer, yet recall from the previous section that nonrigid objects cause explosive hits with long hit-to-ring gaps. Nonrigid hits create rings with a delay, and yet diminish rings without delay. And, similarly, rigid hits create rings without delay, but are slow dampeners of rings.

These gong observations are crucial for understanding what happens to voiced and unvoiced plosives when they are not released (i.e., when the air in the mouth and lungs is not allowed to burst out, creating the explosive hit sound), which often occurs at word endings (as discussed in the section titled "Two-Hit Wonder"). When a plosive is not released, there clearly cannot *be* a hit-to-ring gap—because it never rings. So how do voiced and unvoiced plosives retain their voiced-versus-unvoiced distinction at word endings? For example, consider the word "bad." How do we know it is a "d" and not a "t" at the end, given that it is unreleased, and thus there is no hit-to-ring delay characterizing it as a "d" and not a "t"?

My gong story makes a prediction in this regard. If voiced plosives really have their foundation in rigid objects (mimicking rigidity's imperceptibly tiny hit-to-ring gap at a word's beginning), then, because rigid objects are poor mufflers, the sonorant preceding an unreleased *voiced* plosive at a word *ending* should last longer than the sonorant preceding an unreleased *un*voiced plosive at a word ending. For example, the vowel sound in "bad" should last longer than in the word "bat." The nonrigid "t" at the end of the latter should muffle it quickly. Are words like "bad" spoken with vowels that ring longer than in words like "bat"?

Yes. Say "bad" and "bat." The main difference is *not* whether the final plosive is voiced—neither is, because neither is ever released, and thus neither ever gets to ring. Notice how when you say "bad," the "a" gets more drawn out, lasting longer, than the "a" sound in "bat." Most non-linguist readers may never have noticed that the principal distinguishing feature of voiced and unvoiced plosives at word endings is not whether they are voiced at all. It is a seemingly unrelated feature: how long the preceding vowel lasts. But, as we see from the physics of events, a longer-lasting ring before a dampening hit *is* the signature of a rigid object's bouncy hit, and so there is a fundamental ecological order to the seemingly arbitrary linguistic phonological regularity. (See Figure 8.)

| HIT (PLOSIVE) | RIGID (VOICED) | NON-RIGID (UNVOICED) |
|---|---|---|
| INITIATES RING (RELEASED) | Short hit-to-ring delay (Short voice-onset time; "da") | Long hit-to-ring delay (Long voice onset time, "ta" |
| MUFFLES RING (Unreleased) | Slow to dampen ring (Preceding sonorant has long duration; "bad") | Quick to dampen ring (Preciding sonorant has short duration; "bat") |

FIGURE 8. Matrix illustrating the tight match between the qualities of hits (not in parentheses) and plosives (within parentheses). For hits, the columns distinguish between rigid and non-rigid hits, and the rows distinguish between hits that initiate rings and hits that muffle rings. Inside the matrix are short descriptions of the auditory signature of the four kinds of hit. For plosives, the columns distinguish the analogs of rigid and non-rigid hits, which are, respectively, voiced and unvoiced plosives; the rows distinguish the analogs of ring-initiating and ringmuffling hits, which are, respectively, released and unreleased plosives. Together, this means four kinds of hit, and four expected kinds of plosive, matching the signature features of the respective

hits. If the meaning of voiced versus unvoiced concerns rigid versus non-rigid objects, then we expect that plosives at word-starts should have little to a lot of voice-onset time, respectively, for voiced and unvoiced. And we expect that for plosives at word endings the voiced ones should reveal themselves via a longer preceding sonorant (slow to damp) whereas unvoiced via a shorter preceding sonorant (fast to damp). Plosives do, in fact, modulate across this matrix as predicted from the ecological regularities of rigid and non-rigid hits at ring-inceptions and ringdampenings.

Over the last half dozen sections of this chapter we have analyzed the constituents—the hits, slides and rings—of events and language. Hits, slides, and rings may be the fundamental building blocks for human speech, but that alone doesn't make speech sound natural. Just as natural contours can be combined in unnatural ways for vision, natural sound atoms can be combined unnaturally for audition. Language will not effectively harness our auditory system if speech combines plosives, fricatives, and sonorants in unnatural ways, like "yowoweelor" or "ptskf". To find out whether speech sounds like nature, we need to understand how nature's phonemes combine, and then see if language combines in the same way. For the rest of this chapter, we will look at successively larger combinations of sounds. But we turn first to the simplest combination.

## NATURE'S SYLLABLES

My friend's boy made a video of himself solving a Rubik's Cube blindfolded, and then posted it on the Web. As I watched him put the blindfold on, pick up the cube, and begin twisting, I noticed something strange about the sound, but I couldn't put my finger on what was unusual. Later, when I commented to my friend how his bright boy must owe it to inheritance, he replied, "Indeed, the apple doesn't fall far from the tree. He faked it. The movie was in reverse."

The world does not sound the same when run backward. What had raised my antennae when watching the Rubik's Cube video was the unusual sounds that occur when one hears events in reverse. One of the first strange sounds occurred when he picked up the cube at the start of the video. Knowing now that it was shown in reverse, what appeared in the video to be him picking up the cube to begin unscrambling it was *actually* him setting the cube down after having scrambled it. Setting the cube down caused a hit and a ring, but in reverse what one hears is a ring coming out of nowhere, and ending with a sudden ring-stopping hit (the second voice of a hit, as discussed earlier in the section titled "Two-Hit Wonder"). That just doesn't happen much in nature. When nature comes to the door, it knocks before ringing, not the other way around. Rings don't *start* events. Rings are due to the periodic vibrations of objects, and objects do not typically ring without first being in physical contact with another object. Rings therefore do not typically occur without a hit or slide occurring first.

Hits, slides, and rings may be the principal fundamental building blocks for events, but rings are a different animal than hits and slides. Hits and slides involve objects in motion, physically interacting with other objects. Hits and slides are the backbone of the causal chain in an event. Rings, on the other hand, occur as a result of hits or slides, but don't themselves cause more events. Rings are free riders, contributing nothing to the causality. Events do not have a ring followed by another ring. That's impossible (although a single complex, or wiggly, ring is possible, as we discussed in an earlier section). And events never have an interaction (i.e., a hit or a slide) followed directly by another interaction without an intervening ring. Sometimes a ring will be inaudible, and so there will *appear* to be two interactions without an intervening ring, but physically there's always an intervening ring, because objects that are involved in a physical interaction always vibrate to some extent. Events also always end with a ring, although whether it is audible is another matter.

The most basic way in which hits, slides, and rings combine is, then, this:

Interaction – Ring

where the interaction can be either a hit or a slide. If we let *c* stand for a hit or a slide (because "c" can be pronounced either as a plosive, "k," or as a fricative, "s"), and *a* stand for a ring (which, recall, can sometimes be wiggly), then the fundamental structure of solid-object physical events is exemplified by *caca*. Not *acac*. Not *cccaccca*. Not *accacc*. And so on. Letting *b* stand for hits and *s* for slides, events take forms such as *ba, sa, baba, saba, basaba*, and so on. Not *ab* or *sba* or *a* or *bbb* or *ssb* or *assb* or the like. This interaction-ring combination is perhaps *the* most fundamental event regularity in nature, and is perhaps the most perceptually salient. Objects percussively interact via either a hit or slide, and give off a ring. Our auditory system—and probably that of most other mammals—is designed to expect Nature's phonemes to come in this interaction-ring form.

Given the fundamental status of interaction-ring combinations, if language harnesses the innate powers of our auditory system, then we expect language to be built out of vocalizations that sound like interaction-ring. Do languages have this feature? That is, do plosives and fricatives tend to be followed by sonorants? Yes. A plosive or fricative followed by a sonorant is, in fact, the most basic and most common phoneme combination across languages. It is the quintessential example of a *syllable*. Words across humankind tend to look approximately like *ca*, or *caca*, or *cacaca*, where *c* stands for a plosive or fricative, and *a* for one or more consecutive sonorants. All languages have syllables of this *ca* form. And many languages— such as Japanese—*only* allow syllables of this form.

Whereas interaction-ring is the most fundamental natural combination of event atoms, ring-interaction is a combination that is *not* possible. A ring followed by an interaction sounds out of this world, as in my friend's son's Rubik's Cube video. We therefore expect that languages tend to avoid combinations like *ac* and *acac.* This is, in fact, the case. The rarest syllable type is of this *ac* form, and words starting with a sonorant and followed by a plosive or fricative are rare. In data I collected at RPI in 2008 with the help of undergraduate student Elizabeth Counterman and graduate student Kyle McDonald, about 80 percent of our sampled words (with three or fewer non-sonorants) across 18 widely varying languages begin with a plosive or a fricative. (See the legend of Figure 12 for a list of the sampled languages.) And a large proportion of the words starting with a sonorant start with a nasal, like "m" and "n", the least sonorant-like of the sonorant consonants (nasals at word starts can have a fairly sudden start, and are more plosive-like than other sonorant consonants).

Note that a word starting with a vowel does *not* start with a sonorant, because when one speaks such a word, the utterance actually begins with something called a *glottal plosive*, produced via the sudden hit-like release of air at one's voice box. To illustrate the glottal plosive, slowly say "packet," and then slowly say "pack it." When you say the latter, there can often be a sharp beginning to the "it," something that will never occur before the "et" sound in "packet." That sharp beginning is the glottal plosive. Words starting with sonorants are, thus, less common than one might at first suspect. Even words like "ear," "I," "owe," and "owl," then, are cases of plosives followed by sonorants, and agree with the common hit-ring (the most common kind of interaction-ring) structure of nature.

Words truly beginning with a sonorant sound begin not with a vowel, but with a sonorant consonant like w, y, l, r and m. When one says, "what," "yup," "lid," "rip," and "map," the start of the word is non-sudden (or less sudden than a plosive), ramping up more gradually to the sonorant sound instead. And notice that words such as these—with a sonorant at the start and a plosive at the end—*do* sound like backwards sounds. Try saying the following meaningless sentence: "Rout yab rallod." Now say this one: "Cort kabe pullod." Although they are similar, the first of these meaningless sentences sounds more like events in reverse. This is because it has words of the ring-hit form, the signature sound of a world in reverse. The second sentence, while equally meaningless, sounds like typical speech (and event) sounds, because it starts with plosives.

Language's most universal structure above the level of phonemes—the syllable—has its foundation, then, in physics. The interaction-rings of physical events got instilled into our auditory systems over hundreds of millions of years of vertebrate and mammalian evolution, and culture shaped language to sound like physics in order to best harness our hardware.

Before we move next to the shape of words, there is another place where syllables play a central role: in rhyme. Two words rhyme if their final syllables have the same sonorant sound, *and* the same plosive or fricative following the sonorant—for example, "snug as a bug in a rug." The sonorant sound is the more important of the two: "bug" rhymes better with "bud" than with "bag." Our ecological understanding of syllables may help to make sense of the perceptual salience of rhyme. When two events share the same ring sound, it means the same kind of object is involved in both events. For example, "tell and "sell" rhyme, and in terms of nature's physics, they sound like two distinct events involving the same object. "Tell" might suggest that some object has been hit, and "sell" that that same object is now sliding. The "ell" in each case signals that it is the same object undergoing different events. This is just the kind of gestalt perceptual mechanism humans are well known to possess: we attempt to group stimuli into meaningful units. In vision this can lead to contours at distant corners of an image being perceptually treated as parts of one and the same object, and in audition it can lead to sounds separated by time as nevertheless grouped into the same object. *That's* what happens in rhyme: the second word of a rhyming pair may occur several lines later, but our brain hears the similar ringing sound and groups it with the earlier one, because it would be likely in nature that such sounds were made by one and the same object.

## IN THE BEGINNING

The Big Bang is the ultimate event, and even *it* illustrates the typical physical structure of events: it started with a sudden explosion, one whose ringing is still "heard" today as the background microwave radiation permeating all space. Slides didn't make an appearance in our universe until long after the Bang. As we will see in this section, hits, slide and rings tend to inhabit different parts of events, with hits and rings—bangs—favoring the early parts.

To get a feeling for where hits, slides, and rings occur in events, let's take a look at a simpler event than the one that created the universe. Take a pen and throw it onto a table. What happened? The first thing that happened is that the pen hit the table; the audible event starts with a hit. Might this be a general feature of solid-object physical events? There are fundamental reasons for thinking so, something we discussed in the earlier section, "Nature's Other Phoneme." We concluded that whereas hits can occur without a slide preceding, slides do *not* tend to occur without a preceding hit. Another reason why slides do not tend to start events is that friction turns kinetic energy into heat, decreasing the chance for the slide to initiate much of an event at all. So, while hits can happen at any part of an event, they are most likely to occur at the start. And while slides can also happen anywhere in an event, they are less likely to occur near the start. Note that I am not concluding that slides are more common than hits at the non-starts. Hits are more common than slides, no matter where one looks within solid-object physical events. I'm only saying that hits are more common at event starts than they are at non-starts, and that slides are *less* common at event starts than they are at non-starts.

Is this regularity about the kinds of interaction at the starts and non-starts of events found in spoken language? Yes. Words of the form *bas* are more common than words of the form *sab* (where, as earlier, *b* stands for a plosive, *s* for a fricative, and *a* for any number of consecutive sonorants). Figure 9 shows the probability that a non-sonorant is a plosive (rather than a fricative) as one moves from the start of a word to non-sonorants further into the word. The data come from 18 widely varying languages, listed in the legend. One can see that the probability that the non-sonorant phoneme is a plosive begins high at the start of words, after which it falls, matching the pattern expected from physics. And, as anticipated, one can also see that the probability of plosives after the start is still higher than the probability of a fricative.

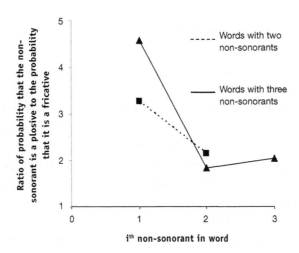

**FIGURE 9.** This shows how plosives are more probable at the start of words, and fall in probability after the start. The yaxis shows the plosive-to-fricative ratio, and the x-axis the i[th] non-sonorant in a word. The dotted line is for words with two non-sonorants, and the solid line for words with three non-sonorants. The main points are **(i)** that plosives are always more probable than fricatives, as seen here because the plosive-to-fricative probability ratios are always greater than 1, and **(ii)** that the ratio falls after the start of the word, meaning fricatives are disproportionately rare at word starts. These data come from common words (typically about a thousand) from each of the following languages: Japanese, Zulu, Malagasy, Somali, Fijian, Lango, Inuktitut, Bosnian, Spanish, Turkish, English, German, Bengali, Yucatec, Wolof, Tamil, Taino, Haya.

We just concluded that hits are disproportionately common at the starts of events in nature, and that this feature is also found in language. But we ignored rings. Where in events do rings tend to reside? In the previous section ("Nature's Syllables") we discussed the fact that rings do not start events, a phenomenon also reflected in language. How about after the start of a word? There would appear to be a simple answer: rings always occur after physical interactions, and so rings should appear at *all* spots within events, following each hit or slide.

But as we will see next, reality is more subtle.

## THE FIRST WAS A DOOZY

While it is true that all physical interactions cause ringing, the ringing need not be audible, a point that already came up in the section called "Two-Hit Wonder." In this light, we need to ask, *where in events are the rings most audible?* Consider the generic pen-on-table event again. The beginning of that event—the audible portion of it, starting when the pen hit the table—is where the greatest energy tends to be, and the ring sound after the first hit will therefore tend to be the loudest. If the pen bounces and hits the table again, the ring sound will be significantly lower in magnitude, and it will be lower still for any further bounces. Because energy tends to get dissipated during the course of an event, rings have a tendency to be louder earlier in the event than later in the event. This is a tendency, but it is not always the case. If energy gets added during the event, ring magnitude can increase. For example, if your pen bounces a couple of times on the table, but then bounces *off* the table onto the floor, then the floor hit may well be louder than the first table hit (gravity is the energy-adding culprit). Nevertheless, in the generic or typical case energy will dissipate over the course of a physical event, and thus ringing magnitude will tend to be reduced as an event unfolds. Therefore, the audibility of a ring tends to be higher near the start of an event; or, correspondingly, the probability is higher later in an event that a ring might *not* be audible.

If language is instilled with physics, we would accordingly expect that sonorant phonemes are more likely to follow a plosive or fricative near the start of a word, and are more likely to go missing near the end of a word. This is, in fact, the case. Figure 10 shows how the probability of a sonorant following a non-sonorant falls as one moves further into a word, using the same data set mentioned earlier. For example, words like "pact" are not uncommon in English, but words like "ctap" do not exist, and are rare in languages generally.

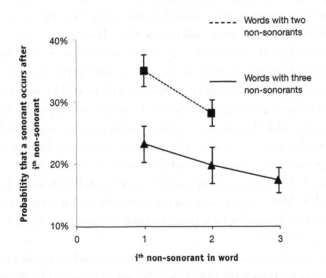

**FIGURE 10.** This shows that sonorant phonemes are more probable near the starts of words, namely just after the first non-sonorant (usually a plosive). The square data are for words having two non-sonorants, and the triangle data for words having three non-sonorants.

We have begun to get a grip on how hits, slides and rings occur within events, but we have only considered their probability as a function of how far into the event they occur. In real events there will be complex dependencies, so that if, say, a slide occurs, it changes the probability of another slide occurring next. In the next section we'll ask, more generally, which combinations of hits and slides are common and which are rare, and then check for the same patterns in language.

## NATURE'S WORDS

Rube Goldberg machines excel at producing very long events, all part of a single causal chain. Like most events, Rube Goldberg events are built mostly out of hits, slides, and rings. Again letting *b*, *s*, and *a* stand for hits, slides and rings, Rube Goldberg events sound something like *basabababasa-babababababasababababasa*, although the chains are very often much longer than even this. If events were typically like Rube Goldberg events, then even if spoken words have many of the auditory features found in events, words would be much too short to be event-like. Events are, however, not typically Rube Goldberg-like. Events are, instead, much more typically like a pen thrown on a table, the generic event we discussed in the previous section. Pen-on-table events may consist of a hit, hit, and slide. Or possibly just a hit and a slide. Or even just a lone hit. Most events have just several physical interactions or fewer, much nearer in length to spoken words than to Rube Goldberg events.

This is what nature-harnessing expects. Spoken words across human languages are not only built out of sounds like those in solid-object physical events, words tend to have the *size* of typical physical events. Words tend to sound like events with up to several interaction sounds—plosives or fricatives—not, say, ten. And although words with a single interaction sound are allowed, two or three interaction sounds are more common, again like solid-object physical events.

Words are not only approximately the size of solid-object physical events—i.e., having several interaction sounds— words also take the amount of *time* for a typical event. This is something I have thus far ignored. But notice that plosives, fricatives, and rings do not just have similar acoustic characteristics to hits, slides and rings; they also occur over periods of time similar to those typical of hits, slides, and rings. For example, although I described both hits and plosives as nearly instantaneous explosions, the notion of "instantaneous" depends on the time scale relevant to the listener—what's instantaneous to a human may not be instantaneous to a fly. Hits and plosives are both instantaneous explosions as heard by *human* ears. This is why plosives sound hit-like; for example, if a hit-like sound were stretched out it would, instead, sound more slide-like (something we discussed in the earlier section called "Hesitant Hits"). Similarly, fricatives and sonorants tend to occur over time scales similar to the slides and rings of physical events. Typical syllables of human speech—e.g., of the form *ba* or *sa*—tend to have a duration approximately on the order of tenths of seconds, roughly the same time scale as is common for physical events involving macroscopic objects. In fact, you'll notice in Figure 4 back in Chapter 1 that the physical and linguistic analogs (e.g., a hit and "k") are on the same scale for the time (x) axis.

Words tend to be built out of the constituents of natural solid-object physical events, and to have approximately the size and temporal duration of such events. But are words actually *structured* like solid-object physical events? Are the natural-sounding phonemes and syllables put together into natural-sounding words? In particular, I'm interested in asking whether the sequences of physical interactions that occur in events—the hits and slides—are similar to the sequences of plosives and fricatives in words. My students and I analyzed the "event structure" of common words across 18 languages, and for each language we measured the distribution of six event types: hit (*b*), slide (*s*), hit-hit (*bb*), hit-slide (*bs*), slide-hit (*sb*), and slide-slide (*ss*). For example, "tea" is a b, "far" is an s, and "faker" is an sb.

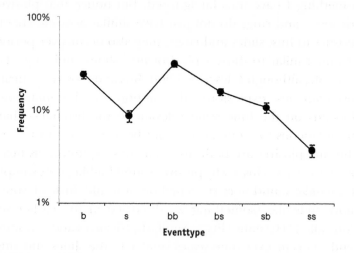

**FIGURE 11.** The freqency of the structure types found in words across 18 widely diverse languages (listed in the legend of Figure 12). (Standard error bars shown. See Appendix B for details.)

To estimate how common these simple event types are in nature, students Elizabeth Counterman, Kyle McDonald, and Romann Weber counted the kinds of events occurring in a wide variety of videos. In deciding upon the kinds of videos to sample, we were not especially interested in having videos of, say, the savanna. Recall our discussion in the previous chapter, where we observed that there are "hard cores" of nature likely found in most or all habitats with solid objects crashing about. In choosing twenty videos from which to enumerate solid-object physical events, we simply aimed for a variety of scenarios in which solid-object physical events occur, including cooking, children playing, family gatherings, assembly instructions, and acrobatics. Each student acquired data on the events occurring, and did so using only the visual modality (that is, the videos were on mute); this helped to deal with a worry that our auditory systems are biased by speech so that we hear speech-like structure in events (akin to seeing faces in clouds). The three observers identified an average of 650 events across the twenty videos. Figure 12 shows the average results for the videos as a dotted line, overlaid on the language data from Figure 11. One can see the close similarity in the plots. (Notice that a simple model assuming hits are more common than slides does not explain why bs occurs more often than sb in the language data.)

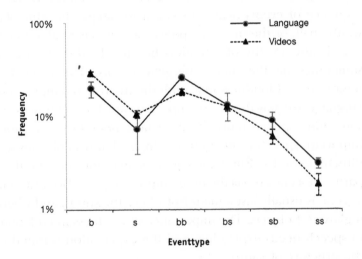

**FIGURE 12.** The relative frequency of simple event types in videos and in language. One can see their considerable similarity. (Standard error bars shown. See Appendix B for details.)

Again, we find the signature of solid-object physical events—of nature—in spoken language! Our final story in this chapter on speech concerns the sounds of speech above the level of words: the structure of whole phrases and sentences.

## UNRESOLVED QUESTIONS

Earlier in the chapter I remarked on how audition is nature's more terse modality, only speaking up when there's an event. In real life, though, there can often be "event overload." I'm sitting at an airport right now, and I just counted 30 distinct sound events occurring around me over the last 30 seconds. How can we possibly pick out the sounds that matter to us amongst all the noise? There *are*, in fact, auditory cues that can tell an observer whether an event is relevant to him or her. In particular, these cues can tell the observer that "an event you should pay attention to is coming."

The most obvious such auditory cue is loudness. As a sequence of events nears me—be it footsteps, the whir of a whiffle ball, or the siren of a police car—it gets louder. Loudness is also worthy of attention because louder events can sometimes be the more energetic events. The ecological importance of loudness may underlie the role of emphasis in language, the way that more important words or sentences are sometimes spoken more loudly. That louder speech is more important speech is one of those things that is so obvious it is difficult to notice. But its analog in vision is not true—brighter parts of a scene are *not* the more important parts. Brightness in a scene is usually just a matter of where the sun is, and where it glares off objects. The importance of loudness modulations in speech needs explaining, and the explanation is found in the structure of nature.

In addition to loudness, events in nature have another sound quality that is even more informative: pitch (the musical, note-like quality of sound). The pitch of an event depends not on how close it is to the observer, but on the *rate* at which it is getting closer to the observer. To understand why, let's imagine standing next to a passing train, the standard example used to explain the Doppler effect. The main observation is that the pitch of the train's whistle starts high and changes to low as it passes. More specifically, note that when the train is far away and approaching, its whistle is at a fixed high pitch, that is, a pitch that is not changing. (It is actually falling, but negligibly and imperceptibly.) The pitch only begins falling audibly when the train is very close to passing you. And shortly after the train has passed you, the pitch has dropped to nearly its low point, so that from then on the pitch stays effectively constant and low. This drop in pitch would apply in any scenario where sequences of events are passing us by. It also occurs any time *we* are moving past noisy objects. Our auditory systems can sense pitch changes on the order of half a percentage of the sound frequency, sufficient for sensing (if not consciously) the pitch changes due to our walking by a source of sound.

The important conclusion of these observations is that a typical sequence of events will tend to have this signature *falling* pitch (unless headed *directly* toward you). One might speculate that this is why language has a tendency to signal the approaching end of a sentence with a falling intonation—a drop in pitch. *That's* what events typically sound like in nature.

Sequences of events do not *always* have pitches that fall, however. Pitches can sometimes rise, but special circumstances are required. First, let's consider what happens if you stand *on* the railroad tracks rather than beside them. Now the pitch of the train stays the same, right up to the moment that it hits you. Of course, at the instant it hits you, the sound you would be hearing if you were conscious abruptly drops to a lower pitch (because it passes you in a single brain-crushingly short instant), and stays at that pitch as the train moves away. A constant pitch accompanied by increasing loudness is the signature of an impending collision. That same loudness increase, but with a pitch *decrease*, signals a near miss.

What could make a pitch *increase?* Considering the train again, imagine first standing beside the tracks as it approaches, but then walking onto the tracks before it gets there. Because you have moved to a position more directly in the train's path of motion, the frequency your ears receive from the train will increase as you walk onto the tracks. Alternatively, the pitch would also increase if you stayed off to the side, but the train jumped the tracks and headed toward you at the last moment. A pitch increase is the signature of a sequence of events that is changing its direction in *your* direction. This is true not only when an approaching sequence of events veers toward you, but also when a receding sequence of events veers so as to begin turning around, perhaps to come back and get you after a miss. An increase in the pitch is, in a sense, more important than loudness. An event might be loud and getting louder, but if its pitch is decreasing, it is not going to hit you. But if an event is not so loud, but has a pitch that is *increasing*, that means it is aiming itself more toward you (or you are aiming more toward it).

A rising pitch suggests, then, that the sequence of events is not finished. Events are coming your way. Or, if the sequence of events is moving away from you, then a rising pitch means it is beginning to turn around. This unresolved nature of rising pitches may be the reason why rising pitches in many languages tend to indicate a question. The spoken sentence, "Is that the elephant that stepped on your car?" is a request for further speech. And what better way to sound unresolved than to mimic the sound of nature's unresolved events?

This is a natural lead-in to the rest of the book, which deals with the origins of music, where loudness and pitch are even more crucial. We will see that "unresolved" pitch even tends to get resolved in melody.

## SUMMARY TABLE

In our modern lives we hear hits, slides, and rings all around us, and we also hear the sounds of speech. They *mean* fundamentally different things to us, and so our brains quickly learn to treat them differently. Our brains can treat them differently because, despite the many similarities between solid-object physical event sounds and speech sounds that I have pointed to throughout this chapter, there are ample auditory cues distinguishing them (e.g., the timbre of a voice is fundamentally different from the timbre of most solid objects). And once our brains treat these sounds as fundamentally different in their ecological meaning, it can be next to impossible to hear that there *are* deep similarities in how they sound. A fish struggling up onto land for the first time, however, and listening to human speech intermingled with the solid-object event sounds in the terrestrial environment, might find the similarity overwhelming. "What is wrong with these apes," it might wonder, "that they spend so much of their day mimicking the sounds of solid-object physical events?"

In this chapter, I have tried to bring out the fish in all of us, pointing out the solid-object event sounds we make when we're speaking, but fail to notice because of our overfamiliarity with them. The table below summarizes the many ways in which speech sounds like solid-object physical events, with references to the earlier sections where we discussed each of them.

| SECTION OF CHAPTER | SOLID-OBJECT PHYSICAL EVENTS | LANGUAGE |
|---|---|---|
| 1. Mother Nature's Voice | Physical events are best sensed by audition. | Language uses audition. |
| 2. Nature's Phonemes | The main three event constituents are hits, slides and rings. | The main three kinds of phoneme are plosives, fricatives and sonorants. |
| 3. Nature's Phonemes | Hits are more common than slides. | Plosives are more common than fricatives. |
| 4. Wiggly Rings | Rings can change in timbre and tone during their occurrence. | Sonorants can change in formants (diphthongs and sonorant consonants) and tone during utterance. |
| 5. Wiggly Rings | Hits and slides do *not* tend to change their sound during their occurrence. | Plosives and fricatives do *not* tend to change during their utterance. |

| SECTION OF CHAPTER | SOLID-OBJECT PHYSICAL EVENTS | LANGUAGE |
|---|---|---|
| 6. Nature's Other Phoneme | A fourth main constituent of events is the hit-slide. But not slide-hit. | A fourth main phoneme type is the affricate. But there is no "fricative-plosive" phoneme type. |
| 7. Two-Hit Wonder | A hit between two objects can have two distinct auditory consequences. Usually it is an instanta-neous explosive burst, but sometimes it is a sudden dampening. | Plosives of any kind have two forms, explosive and dampened (usually word-final). |
| 8. Slides that Sing | Slides usually occur on non-regular surfaces, but sometimes occur on surfaces with periodic regularities, leading to sound with periodicity (tonality). | Fricatives are more com-monly voiceless, but are still often voiced. Whether or not a fricative is voiced is usually part of the identity of the fricative (as the surface periodicity is part of the identity of a surface). |
| 9. Hesitant Hits | Hits can vary widely in the rigidity of the objects involved, and thus vary widely in the time from first explosion to the ring. This can help to identify the objects involved. | Plosives vary in the duration of time to the following sonorant sound. This is called the voice onset time (VOT), and is part of a phoneme's identity. |
| 10. Rigid Muffler | Rigid hits (which cause short hit-to-ring delays when initiating a ring) are poor dampeners of rings. | Voiced plosives (which have short voice onset times when released) are, when unreleased at word-endings, preceded by longer sonorant sounds. |
| 11. Nature's Syllables | Hits and slides cause (usually audible) rings. | Plosives and fricatives tend to be followed by a sonorant. This is the basic syllable form, consonant-vowel (CV). |
| 12. In the Beginning | Hits tend to start events disproportionately more often than slides do. | Plosives tend to start words disproportionately more often than fricatives do. |
| 13. The First was a Doozy | Rings are more audible early in an event. | Sonorants are more likely to follow a plosive or fricative near the starts of words. |

| SECTION OF CHAPTER | SOLID-OBJECT PHYSICAL EVENTS | LANGUAGE |
|---|---|---|
| **14. Nature's Words** | The number of interactions in an event tends to be from one to several, and the time scale of natural solid-object physical events tends to be on the order of several hundred milliseconds (with a lot of variability). | The number of plosives or fricatives in a word tends to be one to several, and the time scale of its utterance tends to be on the order of several hundred milliseconds (with a lot of variability). |
| **15. Nature's Words** | The combinations of hits and slides that occur in natural solid-object physical events have a characteristic, theoretically comprehensible pattern | The combinations of plosives and fricatives in words of languages have the signature pattern of solid-object physical events. |
| **16. Unresolved Questions** | Events with rising pitch are often due to the Doppler effect, wherein an object is veering more toward the observer; i.e., it is the signature auditory pattern of an event "headed your way." Falling pitch means the object is directing itself less and less toward you. | Phrases with rising intonation tend to connote a question or something that is unresolved, metaphorically akin to an event suddenly being directed toward you. Phrases with falling intonation tend to connate greater resolution, metaphorically akin to an object veering away from you, which you no longer have to deal with. |

This chapter, together with the fourth chapter in *The Vision Revolution*, argues that our linguistic ability, for both speech and writing, may well be due to nature-harnessing, rather than to a built-in "language instinct" or to general learning. Although language is central to our modern human identity, so is art, and it is natural to wonder whether some of humankind's artistic wonders also have their origins in nature-harnessing. The remainder of the book takes up music, arguably the pinnacle of humankind's artistic achievement.

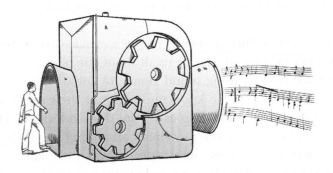

# Soylent Music

## BLIND JOGGERS

Joggers love their headphones. If you ask them why, they'll tell you music keeps them motivated. The right song can transform what is by all rights an arduous half hour of ascetic masochism into an exhilarating whirlwind (or, in my case, into what feels like only 25 minutes of ascetic masochism). Music-driven joggers may be experiencing a pleasurable diversion, but to the other joggers and bikers in their vicinity, they're Tasmanian Devils. In choosing to jog to the beat of someone else's drum rather than their own, headphone-wearing joggers have "blinded" themselves to the sounds of the other movers around them. Headphones don't prevent joggers from deftly navigating the trees, stumps, curbs, and parked cars of the world, because these things can be seen as one approaches them. But when you're moving in a world with other movers, things not currently in front of you can quickly *arrive* in front of you. That's when the headphoned jogger stumbles . . . and crashes into the crossing jogger, passing biker, or first-time tricycler.

These music-blinded movers may be a menace to our streets, but they can serve to educate us all about one of our underappreciated powers: using sound alone, we know where people are around us, and we know the nature of their movement. I'm sitting in a coffee shop as I write this, and when I close my eyes, I can sense the movement all around me: a clop of boots just passed to my right; a person with jingling keys just walked in front of me from my right to my left, and back again; and the pitter-patter of a child just meandered way out in front of me. I sense where they are, their direction of motion, and their speed. I also sense their gait, such as whether they are walking or running. And I can often tell more than this: I can distinguish a brisk from a shuffling walk, an angry stomp from a happy prance; and I can even recognize a complex behavior like turning and stopping to drop a dirty tray in a bin, slowing to open a door, or reversing direction to fetch a forgotten coffee. My auditory system carries out these mover-detection computations even when I'm not consciously attending to them. That's why I'm difficult to sneak up on (although they keep trying!), and why I only rarely find myself saying, "How long has that cheerleading squad been doing jumping jacks behind me?!" That almost never happens to me because my auditory system is keeping track of where people are and roughly what they're doing, even when I'm otherwise occupied.

We can now see why joggers with ears *un*encumbered by headphones almost never crash into feral dogs or runaway grandpas in wheelchairs: they may not see the dog or grandpa, but they *hear* their movement through space, and can dynamically modulate their running to avoid both and be merrily on their way. Without headphones, joggers are highly sensitive to the sounds of cars, and can track their movement: that car is coming around the bend; the one over there is reversing directly toward me; the one above me is falling; and so on. Joggers in headphones, on the other hand, have turned off their movement-detection systems, and should be passed with caution! And although they are a hazard to pedestrians and cyclists, the people they put at greatest risk are themselves. After a collision between a jogger and an automobile, the automobile typically only needs a power wash to the grille.

How does your auditory system serve as a movement-tracking system? In addition to sensing whether a mover is to your left or right, in front or behind, and above or below—a skill that depends on the shape, position and number of ears you have—you possess specialized auditory software that interprets the sounds of movers and generates a good guess as to the nature of the mover's movement through space. Your software has evolved to give you four kinds of information about a mover: (i) his distance from you, (ii) his directedness toward (or away from, or at an angle to) you, (iii) his speed, and (iv) his behavior or gait. How, then, does your auditory system infer these four kinds of information? As we will see in this and the following chapters, (i) distance is gleaned from loudness, (ii) directedness toward you is cued by pitch, (iii) speed is inferred by the number of footsteps per second, and (iv) behavior and gait are read from the pattern and emphasis of footsteps. Four fundamental parameters of human movement, and four kinds of auditory cues: (i) loudness, (ii) sound frequency, (iii) step rate, and (iv) temporal pattern and emphasis. (See Figure 13.) Your auditory system has evolved to track these cues because of the supreme value of knowing what everyone is doing nearby, and where.

This is where things get interesting. Even though joggers without headphones are not listening to music, their auditory systems are listening to fundamentally music-like constituents. Consider the four auditory movement cues mentioned just above (and shown on the right of Figure 13). Loudness? That's just *pianissimo* versus *piano* versus *forte* and so on. (This is called "dynamics" in music, a term I will avoid because it brings

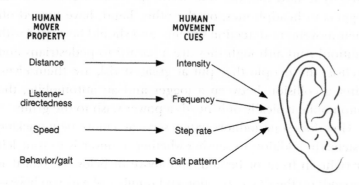

**FIGURE 13.** The four properties of human movers **(left)** are inferred from the four respective auditory stimuli **(right)**.

confusion in the context of a movement theory of music.) Sound frequency? That's roughly pitch. Step rate? That's tempo. And the gait pattern? That's akin to rhythm and beat. The four fundamental auditory cues for movement are, then, mighty similar to (i) loudness, (ii) pitch, (iii) tempo, and (iv) rhythm. (See Figure 14.) These are the most fundamental ingredients of music, and yet, there they are in the sounds of human movers. The most informative sounds of human movers are the fundamental building blocks of music!

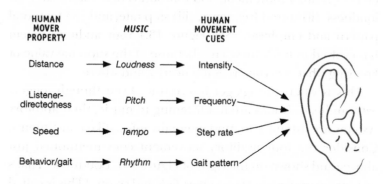

FIGURE 14. Central to music are the four musical properties in the center column, which map directly onto the auditory cues for sensing human movement.

The importance of loudness, pitch, tempo, and rhythm to both music and movement is, as we will see, more than a coincidence. The similarity runs deep—something speculated on ever since the Greeks[*]. Music is not just built with the building blocks of movement, but is actually organized like movement, thereby harnessing our movement-recognition auditory mechanisms. Headphoned joggers, then, don't just miss out on the real movement around them—they pipe fictional movement into their ears a, making them even more hazardous than a jogger wearing earplugs.

Much of the rest of this book is about how music came into the lives of us humans, how it gets into our brains, and why it affects us as it does. In short, we will see that music moves us because it literally sounds like moving.

## THE SECRET INGREDIENT

When I was a teenager, my mother began listening to French instructional programs in order to brush up. She was proud of me when I began sitting and listening with her. "Perhaps my son isn't a square physics kid after all," she thought. And, in fact, I found the experience utterly enthralling. After many months, however, my mother's pride turned to worry, because whenever she attempted to banter in even the most elementary French with me, I would stare back, dumbfounded. "Why isn't this kid learning French?" she fretted.

What I didn't tell my mother was that I wasn't *trying* to learn French. Why was I bothering to listen to a program I could

---

[*] Researchers in this tradition include Alf Gabrielsson, Patrick Shove, Bruno H. Repp, Neil P. McAngus Todd, Henkjan Honing, Jacob Feldman, and Eric F. Clarke (see his *Ways of Listening*).

not comprehend? I will let you in on my secret in a moment, but in the meantime I can tell you what I was *not* listening to it for: the speech sounds. No one would set aside a half hour each day for months in order to listen to unintelligible speech. Foreign speech sounds can pique our curiosity, but we don't go out of our way to hear them. If people loved foreign speech sounds, there would be a market for them; we would set our alarm clocks to blare German at 5:30 a.m., listen to Navajo on the way to work in the car, and put on Bushmen clicks as background for our dinner parties. No. I was not listening to the French program for the speech sounds. Speech doesn't enthrall us—not even in French.

Whereas foreign speech sounds don't make it as a form of entertainment, *music* is quintessentially entertaining. Music *does* get piped into our alarm clocks, car radios, and dinner parties. Music has its own vibrant industry, whereas no one is foolish enough to see a business opportunity in easy-listening foreign speech sounds. And this motivates the following question. Why is music so evocative? Why doesn't music feel like listening to speech sounds, or animal calls, or garbage disposal rumbles? Put simply: why is music *nice* to listen to?

In an effort to answer, let's go back to the French instructional program and my proud, and then concerned, mother. Why *was* I joining my mom each day for a lesson I couldn't comprehend, and had no intention of comprehending? Truth be told, it wasn't an audiotape we were listening to, but a television show. And it wasn't the meaningless-to-me speech sounds that lured me in, but one of the actors. A young French actress, in particular. Her hair, her smile, her mannerisms, her pout . . . but I digress. I wasn't watching for the French language so much as for the French people, one in particular. Sorry, Mom!

What was evocative about the show and kept me wanting more was the *human* element. The most important thing in the lives of our ancestors was the other people around them, and it is on the faces and bodies of other people that we find the most emotionally evocative stimuli. So when one finds a human artifact that is capable of evoking strong feelings, my hunch is that it looks or sounds *human* in some way. This is, I suggest, an important clue to the nature of music.

Let's take a step back from speech and music, and look for a moment at evocative and non-evocative *visual* stimuli in order to see whether evocativeness springs from people. In particular, consider two kinds of visual stimuli, writing and color—each an area of my research covered in my previous book, *The Vision Revolution.*

Writing, I have argued, has culturally evolved over centuries to look like natural objects, and to have the contour structures found in three-dimensional scenes of opaque objects. The nature that underlies writing is, then, "opaque objects in 3-D," and that is *not* a specifically human thing. Writing looks like objects, not humans, and thus only has the evocative power expected of opaque objects: little or none. That's why most writing—like the letters and words on this page—is not emotionally evocative to look at. (See top left of Figure 15.) Colors, on the other hand, are notoriously evocative—people have strong preferences regarding the colors of their clothes, cars, and houses, and we sense strong associations between color and emotions. I have argued in my research and in *The Vision Revolution* that color vision in us primates—our new-to-primates red-green sensitivity in particular—evolved to detect the blood physiology modulations occurring in the skin, which allow us to see color signals indicating emotional state and mood. Color vision in us primates is primarily about the emotions of others. Color is about *humans*, and it is this human connection to color that is the source of color's evocativeness. And although, unlike color, writing is not generally evocative, not all writing is sterile. For example, "V" stimuli have long been recognized as one of the most evocative geometrical shapes for warning symbols. But notice that "V" stimuli are reminiscent of (exaggerations of) "angry eyebrows" on angry faces. Color is "about" human skin and emotion, and "V" stimuli may be about angry eyebrows—so the emotionality in each one springs from a human source. (See top right of Figure 15.) We see, then, that the non-evocative visual signs look like opaque, not-necessarily-human objects, and the evocative visual signs look like human expressions. I have summarized this in the top row of the table in Figure 15.

|  | NOT EVOCATIVE | EVOCATIVE |
|---|---|---|
| VISUAL | Looks like opaque objects in 3D [writing] | Looks like human expression [colors, "V"s] |
| AUDITORY | Sounds like solid-object events [speech] | Sounds like human movement [music] |
|  | "PHYSICS" | HUMAN |

FIGURE 15. Evocative stimuli (right column) are usually made with people, whereas non-evocative stimuli (left column) are more physics-related and sterile.

Do we find that evocativeness springs from the same human source within the auditory domain? Let's start with speech. As we discussed in the previous chapter, speech sounds like solid-object physical events. "Solid-object physical events" amount to a sterile physics category of sound, akin in nerdiness to "three-dimensional world of opaque objects." We are capable of mimicking lots of nonhuman sounds, and speech, then, amounts to yet another mimicry of this kind. Ironically, human speech does not sound *human* at all. It is consequently *not* evocative.

(See the bottom left square of Figure 15 for speech's place in the table.) Which brings us back to music, the other major kind of auditory stimulus people produce besides speech. Just as color is evocative but writing is not, music is evocative but speech sounds are not. This suggests that, just as color gets its emotionality from people, perhaps music gets *its* emotionality from people. Could it be that music, like Soylent Green, is made out of people? (Music has been placed at the bottom right of the table in Figure 15.)

If we believe that music sounds like people, then we greatly reduce the range of worldly sounds music may be mimicking. That amounts to progress: music is probably mostly not about birdsong, wind, water, math, and so on. But, unfortunately, humans make a wide variety of sounds, some in fundamentally different categories, such as speech, coughs, sneezes, laughter, heartbeats, chewing, walking, hammering, and so on. We'll need a more specific theory than one that simply says music is made from people. Next, though, we ask why there isn't any purely visual domain that is as exciting to us as music.

## GOING SOLO

If the visual system and auditory system had competitive streaks, they might argue about which modality has the most compelling art. Each would be allowed to cite as examples only cases exclusively within its own modality: vision-only versus audition-only. This is a difficult contest to officiate. Should vision be allowed to cite all the features of visual design found in culture, such as clothes, cars, buildings, and everyday objects? If so, it would have a big leg up on audition, which is not nearly so involved in the design of our physical artifacts. Let's agree not to include these, by virtue of an "official rule" that the art must be purchased by people for the purpose merely of enjoying the aesthetics, with no other functional benefit. That is, is it vision or audition that commands the greatest portion of the market for art and entertainment?

If you set it up in this way, audition trounces vision. Although the visual modality is found in huge markets like television, videogames, and movies, these rely on audition as well. People put visual art on their walls, but that typically amounts to just a few purchases, whereas it is common to find people who own *thousands* of music albums. The market for the purely visual arts is miniscule compared to that for audition. This is counterintuitive, because if you ask most of us to name the most beautiful things we know of, we are likely to respond with a list of visuals. But when we vote with our pocketbooks, audition wins the solo artist contest. Why is that?

One possible explanation is simply that it is easier to carry on with the chores of life while music is in the background, whereas the visual arts inherently get in the way. Try driving or working or throwing a dinner party while admiring the Mona Lisa. But I suspect it is more than this. If it was merely because of the difficulty of enjoying visual arts while having a life, one might expect us to *want* to stare at beautiful visual art all day, if only we had nothing else pressing to do. Most of us, however, don't exactly fancy the idea of watching visual images all day (without sound). Listening to music all day, however, sounds quite charming! And, in fact, many of us *do* spend our days listening to music.

The stark inequality of vision and audition in this competition for "best solo performer" in the arts is due to a fundamental ecological asymmetry. When we see things in the world, those things are typically making noise. Seeing without hearing therefore feels strange, unnatural, or as if it is missing something. But hearing without seeing is commonplace, because we hear all sorts of things we cannot see—when our eyes are closed, when the source is behind us, when the source is occluded, or when the environment is dark. Sights nearly always come with sounds, but sounds very commonly come without sights. And that's why audition is happy to be a solo artist, but vision isn't. Music is the single-modality artist extraordinaire.

While we now have some idea why there's no solely visual art that rivals music, we still have barely begun our quest to understand why music is so compelling that we are *willing* to purchase thousands of albums.

## AT THE HEART OF A THEORY OF MUSIC

If music sounds human in fundamental respects, as our discussion in the section before last suggested, then it seems to have made heroic efforts to obfuscate this fact. I readily admit that music doesn't sound human to *me*—not *consciously*, at least. But recall the section titled "Below the Radar" from Chapter 1, where I said that we don't necessarily expect cultural artifacts to mimic nature "all the way up." It may be the case that much of our lower-level auditory apparatus thinks that music sounds like humans, but that because of certain high-level dissimilarities, *we*—our conscious selves—don't notice it. How, then, can I hope to convince anyone? I have to convince *you*, after all, not your lower-level auditory areas!

What we need are some qualifying hurdles that a theory of music should have to leap over to gain a hearing . . . hurdles that, once cleared,, will serve to persuade some of Earth's teeming music buffs that music does indeed sound like people moving. Toward this end, here are four such hurdles—questions that any aspiring theory of music might hope to answer.

*Brain*: Why do we have a brain for music?
*Emotion*: Why is music emotionally evocative?
*Dance*: Why do we dance?
*Structure*: Why is music organized the way it is?

If a theory can answer all four questions, then I believe we should start paying attention.

To help clarify what I mean by these questions, let's run through them in the context of a particular lay theory of music: the "heartbeat" theory. Although there is probably more than just one heartbeat theory held by laypeople, the main theme appears to be that a heart has a beat, as music does. Of course, we don't typically hear our own heartbeat, much less others', so when the theory is fleshed out, it is often suggested that the fundamental beat was laid down when we were in utero. One of the constants of the good fetal life was Momma's heartbeat, and music takes us back to those oceanic, one-with-the-universe feelings we long ago lost. I'm not suggesting that this is a good theory, by any means, but it will aid me in illustrating the four hurdles. I would be hesitant, by the way, to call this "lub-dub" theory of music crazy—our understanding of the origins of music is so woeful that any non-spooky theory is worth a look. Let's see how lub-dubs fare with our four hurdles for a theory of music.

The first hurdle was this: "Why do we have a brain for music?" That is, why are our brains capable of processing music? For example, fax machines are designed to process the auditory modulations occurring in fax machine communication, but to *our* ears fax machines sound like a fairly continuous *screech-brrr*—we don't have brains capable of processing fax machine sounds. Music may well sound homogeneously *screechy-brrrey* to non-human ears, but it sounds richly dynamic and structured to *our* ears. How might the lub-dub theorist explain why we have a brain for music? Best I can figure, the lub-dubber could say that our in-utero days of warmth and comfort get strongly associated to Momma's heartbeat, and the musical beat taps into those associations, bringing back warm fetus feelings. One difficulty for this hypothesis is that learned associations often don't last forever, so why would those Momma's-heartbeat associations be so strong among adults? There are lots of beat-like stimuli outside of the womb: some are nice, some are not nice. Why wouldn't those out-of-the-womb sounds become the dominant associations, with Momma's heartbeat washed away? And if Momma's lub-dubs are, for some reason, not washed away, then why aren't there other in utero experiences that forever stay with us? Why don't we, say, like to wear artificial umbilical cords, thereby evoking recollections of the womb? And why, at any rate, do we think we were so happy in the womb? Maybe those days, supposing they leave any trace at all, are associated with nothing whatsoever. (Or perhaps with horror.) The lub-dub theory of music does not have a plausible story for why we have a brain ready and eager to soak up a beat.

The lub-dub theory of music origins also comes up short in the second major demand on a theory of music: that it explain why music is evocative, or emotional. This was the subject of the previous section. Heartbeats *are* made by people, but heartbeat sounds amount to a one-dimensional parameter—faster or slower rate—and are not sufficiently rich to capture much of the range of human emotion. Accordingly, heartbeats won't help much in explaining the range of emotions music can *elicit* in listeners. Psychophysiologists who look for physiological correlates of emotion take a *variety* of measurements (e.g., heart rate, blood pressure, skin conductance), not just one. Heart sounds aren't rich enough to tug at all music's heartstrings.

Heartbeats also fail the "dance" hurdle. The "dance" requirement is that we explain why it is that music should elicit dance. This fundamental fact about music is a *strange* thing for sounds to do. In fact, it is a strange thing for *any* stimulus to do, in *any* modality. For lub-dubs, the difficulty for the dance hurdle is that even if lub-dubs were fondly recalled by us, and even if they managed to elicit a wide range of emotions, we would have no idea why they should provoke post-uterine people to move, given that even fetuses don't move to Momma's heartbeat.

The final requirement of a theory of music is that it must explain the structure of music, a tall order. Lub-dubs *do* have a beat, of course, but heartbeats are far too simple to begin to explain the many other structural regularities found in music. For starters, where is the melody?

Sorry, Mom (again). Thanks for the good times in your uterus, but I'm afraid your heartbeats are not the source of my fascination with music.

Although the lub-dub theory fails the four requirements for a theory of music, the music-sounds-like-human-movement theory of music, as we will see, has answers to all four. We have a brain for music because possessing auditory mechanisms for recognizing what people are doing around us is clearly advantageous. Music is evocative because it sounds like human behaviors, many of which are expressive in their nature—something we will discuss further in a few pages. Music gets us dancing because, as we will also discuss, we social apes are prone to mimic the movements of others. And, finally, the movement theory is sufficiently powerful that it can explain a lot of the structure of music—*that* will require the upcoming chapter and the Encore (at the end of the book) to describe.

## UNDERLYING OVERTONES

The heartbeat theory suffered cardiac arrest, but it was never intended as a serious contender. It was just a prop for illustrating the four hurdles. Speech, on the other hand, is a much more plausible starting point as a foundation for music. But haven't we *already* discussed speech? Wasn't that what the previous chapter was about? We concluded then that speech sounds like solid-object physical events: the structural regularities found among solid-object events are reflected in the phonological patterns of human speech. Speech is all about the phonemes, and how closely they mimic nature's pattern of hits, slides, and ring sounds. Music, on the other hand, cares not a whit for phonemes. Although music can often have words to be sung, music usually gets its identity not from the words, but from the rhythm and tune. Two songs with different words, but with the same rhythm and pitch sequence, are deemed by us to be the same tune, just with different words. That's why we use the phrase, "put words to the music" —because the words (and the phonemes) are not properly part of the meat of the music. The most central auditory feature of speech— its phonological characteristics—is mostly irrelevant to music, making speech an unlikely place to look for the origins of music.

Music is not only missing the phonological core of speech, but it is also missing another fundamental aspect of speech, its most evocative aspect: the meaning, or semantics. If music has its source in speech, and is evocative because of the evocative nature of speech, then why wouldn't music *require* words with meaning, whether metaphorical or direct? Yet, as mentioned above, neither phonology nor words is an essential ingredient of music. (Although phonology and words *are* key ingredients in *poetry*.)

If music comes from speech, then it doesn't come from the phonological patterns of speech, or from the semantics of speech. Although these core functions of speech are dead ends for a theory of music, there is another aspect of speech I have purposely glossed over. People overlay the sterile solid-object event sounds of speech with emotional overtones. We add intonation, a pitch-like property. We vary the emphasis of the words in a sentence, reminiscent of the way rhythm bestows emphasis in music (for instance, the first beat in a measure usually has enhanced emphasis). We vary the timing of the word utterances, akin to the temporal patterns of rhythm in music. And we sometimes modulate the overall loudness of our voices, like a musical crescendo or diminuendo. These prosody-related emotional overtones turn Stephen Hawking computer-voice speech into regular human speech. And these emotional overtones can be understood even in foreign speech, where our ears can often recognize the glib, the mournful, the proud, and the angry. We're just not sure what they are glib, mournful, proud, or angry *about*.

So it is not quite true that speech sounds are sterile. Rather, it is the phonological solid-object event sounds that are sterile. The overtones of speech, on the other hand, are *dripping* with human emotion. Might these overtones underlie music? In an effort to answer, let's discuss the four questions at the heart of any theory of music, the ones I referred to earlier as "brain," "emotion," "dance," and "structure."

Do we have a brain for the overtones of speech? An overtone theory of music would like to say that music "works" on our brains because it taps into speech overtone recognition mechanisms. Are we likely to have neural mechanisms for recognizing overtones of speech? Although I am suggesting in this book that we did not evolve to possess *speech* recognition mechanisms, we primates have been making non-speech vocalizations (cries, laughs, shrieks, growls, moans, sighs, and so on) for tens of millions of years, and surely we have evolved neural mechanisms to recognize them. Perhaps the overtones of speech come from our ancient non-speech vocalizations, and they get laid on top of the solid-object physical event sounds of speech like a whipped cream of evocativeness, a whipped cream our auditory system knows how to taste. An overtone-based theory of music, then, *does* have a plausible story to tell about why our brain would be highly efficient at recognizing overtones.

Can overtones potentially explain the evocativeness of music, the second hurdle we had discussed for any theory of music? Of course! Overtones are *emotional*, used in vocalization to *be* evocative. If music mimics emotional overtones, then it is *easy* to grasp how music can be evocative.

Can an overtone theory of music explain dance, the third hurdle I mentioned earlier? One can see how the emotional non-speech vocalization of other people around us might provoke us into action of some kind—that's probably why people are vocalizing in the first place. That's a start. But we would like to know why hearing overtones would not just tend to provoke us to do stuff, but more specifically, make us move in a time-locked fashion to the emotional vocalizations. I have not been able to fathom any overtone-related story that could explain this, and the absence of any potential connection to dance is a hurdle that an overtone theory stumbles over.

Finally, can overtones explain the structure of music? Do the overtones of speech possess the patterns of pitch, loudness, and rhythm found in music? There is, at least, enough structure floating around in the prosody of speech that one can imagine it might be rich enough to help explain the structure found in music. But despite the nice confluence between ingredients in the overtones of speech and certain similar ingredients in music, overtones appear to be a very different beast from music. First and foremost, what's missing in the overtones of speech is a *beat*, and a rhythm time-locked to a beat. That's the *one* thing the lub-dub theory of music captured, but it is one of the most glaring shortcomings of overtone-based approaches, and it ultimately takes overtones of speech out of the running as a basis for a theory of music.

Before leaving speech for more fertile grounds—in fact, the next section is about sex—consider the two hurdles where overtones appeared promising: "brain" and "emotion." I suggested earlier in this section that overtones could rely on ancient human nonlinguistic vocalizations, but there is another potential foundation for overtones' evocative nature: the sounds of people moving. Rather than music coming from the overtones of speech, perhaps *both* music and overtones have their foundation in the more fundamentally meaningful sound patterns of humans' expressive movements. (And perhaps this is the source of the intersections between music and speech in the brain discerned by Aniruddh D. Patel of the Neurosciences Institute in La Jolla, and other researchers.)

## HOW ABOUT SEX?

Music does not appear to have its origins in the beating heart or in the overtones of speech. That's where I stood on the problem as recently as 2007, when I had recently left Caltech for RPI. I was confident that music was *not* lub-dubs or speech, but I had no idea what music could be. I did, however, have a good idea of some severe constraints any theory of music must satisfy, namely the four hurdles we discussed earlier: brain, emotion, dance, and structure. After racking my brain for some months, and perhaps helped along by the fact that my wife was several months delayed in following me across country to my new job, it struck me: how about sex?

Reputable scientific articles—or perhaps I saw this in one of the women's magazines on my wife's bedside table—indicate that to have sex successfully, satisfying both partners and (if so desired) optimizing the chances of conception, the couple's movements should be in sync with each other. Accordingly, one might imagine that we have been selected to respond to the rhythmic sex sounds of our partner by feeling the urge to match our own movement to his or hers. Evolution would select against people who did not "dance" upon hearing sex moves, and it would also select against people who responded with the sex dance every time a handshake was sufficiently vigorous. The auditory system would thus come to possess mechanisms for accurately detecting the sexual sounds of our partner. A "sex theory of music" of this kind has, then, a story for the "brain" hurdle.

In addition to satisfying the "brain" hurdle, the sex theory also has the beginnings of stories for the other three hurdles. Emotion? Sex concerns hot, steamy bodies, which is, ahem, evocative. Dance? The sex theory explains why we would feel compelled to move to the beat, thereby potentially addressing the "dance" hurdle. (In fact, perhaps the "sex theory" could explain why dance moves are so often packed with sexual over-tones.) And, finally, structure? The sounds of sex often have a beat, the most essential structural feature of music a theory needs to explain.

I was on a roll! But before getting Hugh Hefner on the phone to go over the implications, I needed to figure out how to test the hypothesis. That's simple, I thought. If music sounds like sex, then we should find the signature sounds of sex in music. The question then became, what *are* the signature sounds of sex? What I needed was to collect data from pornography. That, however, would surely land me in a heap of trouble of one kind or another, so I went with the next best thing: anthropology. I began searching for studies of human sexual intercourse, and in particular for "scores" notating the behavior and vocalizations of couples in the act. I also found scores of this kind for nonhuman primates—not my bag— which, I discovered, contain noticeably more instances of "biting" and "baring teeth" than most human encounters. My hope was to find enough of these so that I could compile an average "score" for a sexual encounter, and use it as a predictor of the length, tempo, pitch modulation, loudness modulation, and rhythm modulation of music.

I couldn't find but a handful of such scores, and I did not have the chutzpah to acquire scores of my own. So I gave it up. I could have pushed harder to find data, but it seemed clear to me that, despite its initial promise, sex was far too narrow to possibly explain music. If music sounded like sex, then why isn't all music sexy? And why does music evoke such a wide range of emotions, far beyond those that occur in the heat of sex? And how can the simple rhythmic sounds of sex possibly have enough structure to explain musical structure? Without answers to these questions, it was clear that I would have to take sex off the table.

Enough with the things I don't think can explain music (heartbeats, speech, and sex)! It is about time I begin saying what I think music *does* sound like. And let's edge closer to that by examining what music *looks* like.

## BELIEVE YOUR EYES AND EARWORMS

It is natural to assume that the visual information streaming into our eyes determines the visual perceptions we end up with, and that the auditory information entering our ears determines the events we hear. But the brain is more complicated than this. Visual and auditory information *interact* in the brain, and the brain utilizes both to guess what single scene to render a perception of. For example, the research of Ladan Shams, Yukiyasu Kamitani, and Shinsuke Shimojo at Caltech have shown that we perceive a single flash as a double flash if it is paired with a double beep. And Robert Sekuler and others from Brandeis University have shown that if a sound occurs at the time when the images of two balls pass through each other on a screen, the balls are instead perceived to have collided and reversed direction. These and other results of this kind demonstrate the interconnectedness of visual and auditory information in our brain. Visual ambiguity can be reduced by auditory information, and vice versa. And, generally, both are brought to bear in the brain's attempt to guess about what's out there.

Your brain, then, does not consist of independent visual and auditory systems, with separate troves of visual and auditory knowledge about the world. Instead, vision and audition talk to one another, and there are regions of cortex responsible for making vision and audition fit one another. These regions know about the sounds of looks and the looks of sounds. Because of this, when your brain hears something but cannot see it, your brain does not just sit there and refrain from guessing what it might have looked like. When your auditory system makes sense of something, it will have a tendency to activate visual areas, eliciting imagery of its best guess as to the appearance of the stuff making the sound. For example, when you hear the sound of your neighbor's tree rustling, an image of its swaying, lanky branches may spring to mind. The mewing of your cat heard far away may evoke an image of it stuck high up in that tree. And the pumping of your neighbor's kid's BB gun can bring forth an image of the gun being pointed at Foofy way up there.

Your visual system, then, has strong opinions about the likely look of the things you hear. And, to get back to music, we can *use* the visual system's strong opinions as an aid in gauging music's meaning. In particular, we can ask your visual system what *it* thinks the appropriate visual is for music. If, for example, the visual system responds to music with images of beating hearts, then it would suggest, to my disbelief, that music mimics the sounds of heartbeats. If, instead, the visual system responds with images of pornography, then it would suggest that music sounds like sex. You get the idea.

But to get the visual system to act like an oracle, we need to get it to speak. How are we to know what the visual system thinks music looks like? One approach is to simply ask what visuals are routinely associated with music. For example, when people create imagery of musical notes, what does it look like? One cheap way to find out is simply to do a Google (or any search engine) image search on the term "musical notes." You might think such a search would merely return images of simple notes on the page. However, that is not what one finds. To my surprise, actually, most of the images are like the one in Figure 16, with notes drawn in such a way that they appear to be moving through space. Notes in musical notation don't look anything like this, and *actual* musical notes have no look *at all* (because they are sounds). And yet we humans seem prone to visually depict notes in lively motion.

**FIGURE 16.** Musical notes tend to be visualized like this, a clue to its meaning.

Could these images of notes in motion be due to a more mundane association? Music is played by people, and people have to *move* to play their instruments. Could *this* be the source of the movement-music association? I don't think so, because the movement suggested in these images of notes doesn't *look* anything like an instrument being played. In fact, it is common to show images of an instrument with the notes beginning their movement through space from the instrument: these notes are on their way somewhere, not tied to the musician's key-pressing or back-and-forth swaying.

Could it be that the musical notes are depicted as moving through space because *sound waves* move through space? The difficulty with this hypothesis is that *all* sound moves through space. All sound would, if this were so, be visually rendered as moving through space, but that's not how we portray most sounds. For example, *speech* is not usually visually rendered as moving through space. Another difficulty is that the musical notes in these images are usually meandering, but sound waves don't meander—sound waves go straight. A third problem with the notion that sound waves are the basis for the visual metaphor is that we never *see* sound waves in the first place.

Another possible counter-hypothesis is that musical notes are visually depicted in motion because all auditory stimuli are caused by underlying events that involve movement of some kind. The first difficulty, as with sound waves, is that not all sound, by a long shot, is visually rendered as in motion. The second difficulty is that, while it is true that sounds are typically generated by movement of some kind, it need not be movement of an entire object through space. Moving parts *within* the object may make the noise, without the object going anywhere. In fact, the three examples I gave at the start of this section—leaves rustling, Foofy mewing, and the BB gun pumping—are noises without any bulk movement of the object (the tree, Foofy, or the BB gun, respectively). The musical notes in these images, on the other hand, really do seem to be moving their whole selves across space.

Music is like rustling leaves, Foofy, BB guns, and human speech, in that it is not made by bulk movements through space. And yet music appears uniquely likely to be visually depicted as notes moving through space. And not only moving, but meandering. When visually rendered, music looks alive and in motion (often along the ground)—just what one might expect if music's secret is that it sounds like people moving.

A Google image search on "musical notes" is one way to try to discern what the visual system thinks music looks like. Another is simply to ask ourselves: what is the most common visual display shown during music? That is, if people were to make videos to go with music, what would the videos tend to look like? Luckily for us, people *do* make videos to go with music! They're called music videos, of course. And what do they look like? The answer is so obvious that it hardly seems worth noting: music videos commonly show people moving about, usually in a manner that is time-locked to the music, very often dancing. As obvious as it is that music videos typically show people moving, we must remember to ask ourselves why music isn't typically visually associated with something very different. Why aren't music videos mostly of rivers, avalanches, car races, windblown grass, lions hunting, fire, or bouncing balls? It is because, I am suggesting, our brain thinks that humans moving about is what music *should* look like . . . because it thinks that humans moving about is what music *sounds* like.

Musical notes are rendered as meandering through space. Music videos are built largely from people moving, and in a manner time-locked to the music. That begins to suggest that the visual system is under the impression that music sounds like human movement. But if that's really what the visual system thinks, then it should have more opinions than just "music sounds like movement." It should have opinions about what *kind* of movement music sounds like, and therefore, more exactly what the movement should look like. Do our visual systems have opinions this precise? Are we picky about the visual movement that goes with music?

You *bet* we are! *That's* choreography. It's not OK to play a video of the *Nutcracker* ballet during Beatles music, nor is it OK to play a video of the *Nutcracker* to the music of *Nutcracker*, but with a small time lag between them. Video of human movement has to have all the right moves at the right time to be the right fit for music.

These strong opinions about what music looks like make perfect sense if music mimics human movement sounds. In real life, when people carry out complex behaviors, their visible movements are tightly choreographed with the sounds they make—because the sight and the sound arise from the *same* event. When you hear movement, you expect to *see* that same movement. Music sounds to your brain like human movement, and that's why, when your brain hears music, it expects that any visual of it should match up with it.

We just used your brain's visual system as an oracle to divine the meaning of music, and it answered, "People moving." Let's now use your brain in another oracle-like fashion. If music has been culturally selected to fit the brain, then let's look into which pieces of music are the best fit for the brain, with the idea that these pieces may be the best representatives of what music has been culturally selected to sound like. But how can we gauge which pieces of music are the best fits? One thought is that "symptoms" of a piece of music fitting the brain really well might be that the brain would process it especially easily, remember it easily, and internally hear it easily. Are there pieces of music like this?

Yes, there are! They're called *earworms*—those songs with a tendency to get stuck in people's heads. These pieces of music fit the brain so well that they can sometimes become nuisances. Earworms, then, may be great representatives of the fundamental structural features that have been selected for in music. What are the common qualities of pieces of music that become earworms?

When he was an RPI graduate student, Aaron Fath got interested in this question. He was dissatisfied with the standard line that songs become earworms because they are highly repetitive. *Most* songs are highly repetitive, he reasoned. Instead, he began to notice that a large fraction of earworms have a particular dance or move that goes along with the music. Examples of songs tightly connected to a particular movement include "I'm a Little Teacup," "Macarena," "YMCA," "Chicken Dance," "If You're Happy and You Know It," and "Head, Shoulders, Knees and Toes." Let's call these pieces *movement-explicit*. He also noticed that many other earworms were songs that accompanied specific visual movements (like a commercial jingle on television) or were dance songs (even if no specific movements were associated with them).

Aaron used two existing catalogs of earworms: a top 17 list of earworms from James Kellaris of the University of Cincinnati (obtained by polling 559 students), and a list of "top annoying earworms" from an online poll at the website Keepers of Lists (one user posted 220 songs, and 80 other users voted on whether or not they were earwormy; Aaron took the 38 songs having more than 10 votes). Movement-explicit pieces accounted for 23.5 percent and 18.4 percent of these lists. To gauge whether these are unusually large percentages of movement-explicit pieces, he sampled the #8 song on the Billboard Hot 100 Chart every nine months from 1983 to the present, and among these 38 songs, *none* were of the movement-explicit variety. As a second gauge, he sampled the #1 songs for each year from 1955 through 2006 (defined by Aaron—differently than Billboard does it—as the song released in a year that was #1 on the Billboard Hot 100 for the greatest number of weeks, and thus had the most staying power). Of these 52 songs, only *one* was of the movement-explicit kind (namely, "Macarena").

These data suggest that earworms are disproportionately movement-explicit: about one-fifth of the earworms had specific dance moves that went with them, whereas less than 2 percent of top pop songs are of this kind. Our speculation is that songs become earworms not because they are movement-explicit so much as because they are *consistent with the sounds of people moving*—movement-explicit songs just happen to be under especially strong selection pressure to be consistent with the sounds of people moving. Although only a fifth of the earworms were of the movement-explicit kind, many of the others seemed to be in the "accompaniment" or "dance" category (although we have not yet tried to operationally measure these and compare them to control data sets). An alternative possibility is that when a song becomes tightly linked to movement, it is that very association which helps make it an earworm. This would suggest that music becomes more brain-worthy when packaged together with a motor program, and this, too, would appear to point to the music-is-movement theory.

It *looks* like music may be the sounds of human movement. We asked the expert on how things look: your visual system. Like presenting a deeply encrypted code to an oracle, we asked for the visual system's interpretation of that enigmatic thing called music, and it had a clear and resounding response: music sounds like people moving and doing things, and thus must be visually rendered as humanlike motion in sync with the musical sounds. We also queried your brain in another fashion: we asked it which songs it most revels in, which ones are so earwormalicious that the brain loves to internally sing them over and over again. And the brain answered: the more movement-explicit songs are more likely to be the earwormy ones. The brain seems to be under the impression that music sounds like people moving.

## BRAIN AND EMOTION

The opinion of visual systems and the hints of earworms are interesting and motivating, but we can't just take them at their word. In order to make a solid case that music sounds like human movement, I need to show that the music-is-movement theory can leap the four hurdles we discussed earlier: "brain," "emotion," "dance," and "structure." Let's begin in this section with the first two.

For the "brain" hurdle, I need to say why our brain would have mechanisms for making sense of music and responding to it so eagerly and intricately. For the theory that music sounds like human movement, then, we must ask ourselves if it is plausible that we have brain mechanisms for processing the sounds of humans doing stuff. The answer is yes. *Of course* we have humans-doing-stuff auditory mechanisms! The most important animals in the life of any animal are its conspecifics (other animals of the same species), and so our brains are well equipped to communicate with and "read" our fellow humans. Face recognition is one familiar example, and color vision, with its ability to detect emotional signals on the skin, is another one (which I discussed in detail in my previous book, *The Vision Revolution*). It would be bizarre if we had no specialized auditory mechanisms for sensing the sounds of other people carrying out behaviors. Actions speak louder than words—the sounds we make when we act are often a dead giveaway to what we're up to. And we've been making sounds when we move for many millions of years, plenty long enough to have evolved such mechanisms. The music-sounds-like-movement hypothesis, then, can make a highly plausible case that it satisfies the "brain" hurdle. Our brains surely *have* evolved to possess specialized mechanisms to hear what people are doing.

How about the second hurdle for a theory of music, the one labeled "emotion"? Could the mundane sounds of people moving underlie our love affair with music? As we discussed at the start of the chapter, music is evocative—it can sound joyous, aggressive, melancholy, amorous, tortured, strong, lethargic, and so on. I said then that the evocative nature of music suggests that it must be "made out of people." Human movement *is*, obviously, made of and by people, but can human movement truly be evocative? Of course! The ability to infer emotional states from the bodily movements of others comes via several routes. First and foremost, when people carry out behaviors they move their bodies, movements that can give away what the person is doing; knowing what the person is doing can, in turn, be crucial for understanding the actor's emotion or mood. Second, the actor's emotional state is often cued by its side effects on behavior, such as when an exhausted person staggers. And third, some bodily movements serve as direct emotional signals, more akin to facial expressions and color signals: bodily movements can be proud, strutting, threatening, ebullient, jaunty, sulking, arrogant, inviting, and so on. Human movement can, then, certainly be evocative. And unlike evocative facial expressions and skin color signals, which are silent, our evocative bodily expressions and movements make noises. The sounds of human movement not only are "made from people," then, but they can be truly evocative, fulfilling the "emotion" hurdle.

An example will help to clarify how the sounds of human movement can be emotionally evocative. Michael Zampi, then an undergraduate at RPI. was interested in uncovering the auditory cues for happy, sad, and angry walkers. He first noted that University of Tübingen researchers Claire L. Roether, Lars Omlor, Andrea Christensen and Martin A. Giese had observed that happy walkers tend to lean back and have large arm and leg swings, angry walkers lean forward and have large arm and leg swings, and sad walkers tend to lean forward and have attenuated arm and leg swings.

"What," Michael asked, "are the distinctive *sounds* for those three gaits?" He reasoned that leaning back leads to a larger gap between the sound of the heel and the sound of the toe. And, furthermore, larger arm and leg swings tend to lend greater emphasis to any sounds made by the limbs in between the footsteps (later I will refer to these sounds as "banging ganglies"). Given this, Michael could conclude that happy walkers have long heel-toe gaps and loud between-the-steps gait sounds; angry walkers have short heel-toe temporal gaps and loud between-the-steps gait sounds; and sad walkers have short heel-toe gaps and soft between-the-steps gait sounds. But are these cues sufficient to elicit the perception that a walker is happy, angry, or sad?

Michael created simple rhythms, each with three drum strikes per beat: a toe-strike on the beat, a heel strike just before the beat, and a between-the-step hit on the off-beat. Starting from a baseline audio track—an intermediate heel-toe gap and a between-the-steps sound with intermediate emphasis—Michael created versions with shorter and longer heel-toe gaps, and versions with less emphasized and more emphasized between-the-steps sounds. Listeners were told they would hear the sounds of people walking in various emotional states, and then the listeners were presented with the baseline stimulus, followed by one of the four modulations around it. They were asked to volunteer an emotion term to describe the modulated gait. As can be seen in Figure 17, subjects had a tendency to perceive the simulated walker's emotion accurately.

| | | EMOTION OF TRACK | | |
| --- | --- | --- | --- | --- |
| | | Angry | Happy | Sad |
| | Angry | 9 | 3 | 0 |
| PERCEIVED EMOTION | Happy | 1 | 5 | 1 |
| | Sad | 0 | 2 | 8 |
| | Other | 1 | 1 | 2 |

FIGURE 17. Each column is for one of the three tracks having the sounds modulating around the baseline to indicate the labeled emotion. The numbers show how many subjects volunteered the emotions "angry", "happy", "sad" or other emotions words for each of the three tracks. One can see that the most commonly perceived emotion in each column matches the gait's emotion.

This pilot study of Michael Zampi's is just the barest beginning in our attempts to make sense of the emotional cues in the sounds of people moving. The hope is that by understanding these cues, we can better understand how music modulates emotion, and perhaps why genres differ in their emotional effects.

If music has been culturally selected to sound like human movement, then it is easy to see why we'd have a brain for it, and easy to see why music can be so emotionally moving. But why should music be so *motionally* moving? The music-is-movement theory has to explain why the sounds of people moving should impel other people to move. That's the third hurdle over which we must leap: the "dance" hurdle, which we take up next.

## MOTIONALLY MOVING

Group activities with toddlers are hopeless. Just as you get the top toddler into position at the peak of the toddler pyramid, several on the bottom level have begun crying, pooping, or wandering away. Toddlers prefer to treat their daycare mates as objects to ignore, climb over, or hit. And just try getting a dozen of them to do anything in unison, like performing "the wave" in the audience at a roller derby! If aliens observed us humans only during toddlerhood, they might conclude that we don't get on well in groups, and that, lacking a collaborative spirit, we will be easy prey when they invade.

But brain-thirsty aliens might come to a very different conclusion if they dropped in on a daycare center during music time. Flip on "The Wheels on the Bus Go Round and Round," and a dozen randomly wandering, cantankerous droolers begin shaking their stinky bottoms in unison. Aliens might surmise that music is some kind of marching order, a message from the human commander to activate gyrations against an invading enemy.

Dancing toddlers, of course, play little or no role in explaining why we haven't been invaded by aliens, but they do raise an important question. Why *do* toddlers seem to be compelled to move to the music? And, more generally, why is this a tendency we keep into adulthood? At this very moment of writing, I am, in fact, swaying slightly to Tchaikovsky's Piano Concerto No. 1. Don't I have better things to do? Yes, I do—like write this book. Yet I keep pausing to hear the music, and end up ever so slightly dancing. It is easy to understand why people dance when a gun is fired at their feet like in old Westerns, but music is so much less substantial than lead, and yet it can get us going as surely as a Colt 45. What is the source of music's power to literally move us, like rats to the Pied Piper's flute?

We can make sense of this mystery in light of the theory that music sounds like human movement. If music sounds like movement, and music makes us move, then it is not so much music that is making us move, but the sound of human movement. And *that's* not at all mysterious! *Of course* the behaviors of others may elicit responsive actions from us. For example, if my three-year-old son barrels headlong toward my groin, I quickly move my hands downward for protection. If he throws a rubber ball at my head, I try to catch it. And if he suddenly decides he'd rather not wear his bathing shorts, I quickly pull them back up. Not only do I behave in reaction to my son's behavior, but my behavior must be timed appropriately, lest he career into me, bean me with a ball, or strip buck-naked and get a head start in his dash away. Music sounds like human behavior, and human behavior often elicits appropriately timed behavioral responses in others, so it is not a surprise, in light of the theory, that music elicits appropriately timed behavioral responses.

It's easy to see why three-year-old aggressive and streaking behaviors would prompt a well-timed response in others (especially parents). Another common category of human behavior that elicits a behavioral response in others, in fact one of the most common, is *expressive* behavior. Human expressions are *for* other humans to see or hear or smell, precisely in order to prompt them to modulate their behavior. Sometimes another person's response may be a complex whole-body behavior (I give my wife my come-hither look, she responds by going thither), and sometimes the other person's behavioral response may simply be an expression of emotion (I grimace and rub my newly minted bruise, and my son responds by smiling). If music is good at getting us to move, then, in this light, one suspects that music must usually sound not merely like movement that kicks (literally, in my son's case) listeners into moving in response, but, more specifically, like human emotional or expressive behaviors.

Sound triggering movement. That's starting to sound a bit like dance. To more fully understand dance, we must grasp one further thing: *contagious* behaviors—behavioral expressions that tend to spread (see Figure 11). For example, if I smile, you may smile back; and if I scowl, you'll likely scowl back. Even yawns are catching. And contagious behavior is not confined to the face. Nervous behavior can spread, and angry bodily stances are likely to be reciprocated. If you raise your hands high into the air, a typical toddler will also do so, at which point you have a clear tickle shot. Even complex whole-body behaviors are contagious, accounting for why, for example, people in a crowd often remain passive bystanders when someone is being attacked (other people's inaction spreads), and how a group of people can become a riotous mob (other people's violent behavior spreads). By the way, have you yawned yet?

Music, then, may elicit movement for the same reasons that a cartoon smiley face can elicit smiles in us: music can often sound like *contagious* expressive human behavior and movement, and trigger a similar expressive movement in us. Music may not be marching orders from our commander, but it can sometimes cue our emotional system so precisely that we feel almost compelled to march in lockstep with music's fictional mover. And this is true whether we are adults or toddlers. When music is effective at getting us to mimic the movement *it* mimics, we call it dance music, be it a Strauss waltz or a Grateful Dead flail.

The music-sounds-like-movement theory can, then, explain why music provokes us to dance—the third of the four hurdles a theory of music must leap over. The fourth and final hurdle concerns the *structure* of music, and it will take the upcoming chapter and the Encore chapter to make the case that music has the signature structure of humans moving.

## DON'T ROLL OVER, BEETHOVEN

The case for my theory is strong, I believe, and I hope to convince you that music sounds like human movement. If I am correct, then, with the movement-meaning of music in hand, we will be in a position to create a new generation of "supermusic": music deliberately designed to be even more aesthetically pleasing, by far, than previous generations of music. Music has historically been "trying" to shape itself like expressive human behaviors, in the sense that *that* was what was culturally selected for. But individual composers didn't *know* what music was trying to be—composers didn't know that music works best when tapping into our human-movement auditory mechanisms. Musical works have heretofore tended to be *sloppy* mimickers of human movement. With music decoded, however, we can tune it perfectly for our mental software, and blow our minds. You're toast, Beethoven! I've unraveled your secrets!

No. Just kidding. I'm afraid that the music research I'm describing to you will do no such thing, even if every last claim I make is true. To see why the magic of Beethoven is not unraveled by my theory, consider photographic art. Some photographs have evocative power; they count as art. Some photographs, however, are just photographs, and not art. What exactly distinguishes the art from the "not" is a genuine mystery, and certainly beyond me. But there *is* something that is obviously true about art photographs: they are *photographs*. Although that's obvious to us, imagine for a moment that four-dimensional aliens stumble upon a pile of human artifacts, and that in the pile are photographs. Being four-dimensional creatures, they have poor intuitions about what a three-dimensional world looks like from a particular viewpoint inside it. Consequently, our human photographs are difficult to distinguish from the many other human artifacts that are flat with stuff printed upon them, such as wallpaper, clothing, and money. If they are to realize that the photographs are, in fact, photographs—two-dimensional representations of our 3-D world—they are going to have to *discover* this.

Luckily for them, one alien scientist who has been snooping around these artifacts has an idea. "What if," he hypothesizes, "some of the flat pieces of paper with visual marks are photographs? Not of our 4-D world, but of their human 3-D world?" In an effort to test this idea, he works out what the signature properties of photographs of 3-D worlds would be, such as horizons, vanishing points, projective geometry, field of focus, partial occlusion, and so on. Then he searches among the human artifacts for pieces of paper or fabric having these properties. He can now easily conclude that wallpaper, clothing, and money are not photographs. And when he finds some of our human photographs, he'll be able to establish that they *are* photographs, and convince his colleagues. This alien's research would amount to a big step forward for those aliens interested in understanding our world and how we perceive it. A certain class of flat artifacts is meaningful in a way they had not realized, and now they can begin to look at our photographs in this new light, and see our 3-D world represented in them.

The theory of music I am defending here is akin to the alien's theory that some of those flat artifacts are views of 3-D scenes. To us, photographs are obviously of 3-D scenes; but to the aliens this is not at all obvious. And, similarly, to our auditory system, music quite obviously is about human action; but to our conscious selves this is not in the least obvious (our conscious selves are aliens to music's deeper meaning).

To see why this book cannot answer what is *good* music, consider what this alien scientist's discovery about photographs would *not* have revealed. Unbeknownst to the alien, some of the photographs are considered by us humans to be genuine instances of art, and the rest of the photographs are simply photographs. This alien's technique for distinguishing photographs from non-photographs is no use at all for distinguishing the artful photographs from the mere photographs. Humanity's greatest pieces of photographic art and the most haphazard kitsch would all be in the same bag, labeled "views of a 3-D world." By analogy, the most expressive human movement sounds and the most run-of-the-mill human movement sounds are all treated the same by the ideas I describe in this book; they are all in the same bag, labeled "human movement sounds." Although it is expressive human movements that probably drive the structure of music, I have enough on my hands just trying to make the beginnings of a case that music sounds like human movement. Just as it is easier for the four-dimensional alien to provide evidence for photograph-ness than to provide evidence for artsy-photograph-ness, it is much easier for me to provide evidence that music is human-movement-ish than to provide evidence that it is expressive-human-movement-ish. Photographic art is views of 3-D scenes, but views of 3-D scenes need not be photographic art. Similarly, music is made of the sounds of humans moving, but the sounds of humans moving need not be—and usually are not—music.

Relax, Beethoven—no need to roll over. If the music-sounds-like-movement theory is correct, then it is best viewed as a cipher key for decoding music. It gives our conscious, scientific selves the ability to translate the sounds of music back into the movements of humans (something our own lower-level auditory areas already *know* how to do). But knowing how to *read* the underlying movement meaning of music does not mean one knows how to *write* music. Just as I can read great literature but cannot create it, a successful music-is-movement theory will allow us to read the meaning of music but not to compose it. Creating good music requires knowing which human movements are most expressive, and making music sound like *that*. But a theory of *expressive* human movements is far harder to formulate than a theory of human movements generally. All I can hope to muster is a general theory of the sounds of human movements, and so the theory will be, at best, a decoder ring, not a magical composer of great music.

But a decoder ring may nevertheless be a big step forward for composers. Composers have thus far managed to create great music—great auditory stories of human movements, in our theory's eyes—without explicitly understanding what music means. With a better understanding of the decoder ring, composers can consciously employ it in the creative process. Similarly, the four-dimensional alien has much better odds of mimicking artistic photography once he has figured out what photographs actually look like. Until then, the alien's attempts at artistic photography wouldn't even look like photography. ("Is this photographic art?" the alien asks, holding up a plaid pattern.) The aliens must know what basic visual elements characterize photography before they can take it to the next level, start to guess which arrangements of those elements are superior, and try their own tentacles at art photography. You can't have expressive photography without photography, and you can't have expressive human movement sounds without human movement sounds. The theory of music I'm arguing for, then, does not explain what makes great music. But the theory would nevertheless be a big step forward for this. Like the alien's basic discovery, it will enable us to pose hypotheses about why some music is great—by referring to the expressive movements and behaviors it depicts.

This decoder ring will, then, be helpful to composers, but it cannot substitute for the expressive antennae composers use to create musical art. For choreographers and movie composers, this decoder ring is potentially much more important. Choreographers and movie composers are deeply concerned with the mapping of music to movement (the principal domain of choreography) or from movement to music (the principal domain of movie composers), and so a decoder ring that translates one to the other is a potential holy grail. In reality, though, it's not as simple as that. A given piece of music probably does not determine *particular* dance moves (although your auditory system may pick out just one movement)—a good choreographer needs an artistic head to pick the most appropriately expressive of the many possible movements consistent with the music. And for any given movie visual, a good film composer will have to use his or her artistic talents to find an appropriately expressive theme for the scene. Any music–movement decoding devices made possible by this book won't put choreographers or movie composers out of work, but such a decoder may serve as an especially useful tool for these disciplines, providing new, biologically justified constraints on what makes a good music–movement match.

So, what *is* great music? I don't know. My only claim is that it tends to be written in the language of human movement. Music is movement. But it is not the case that movement is music. Just as most stories are not interesting, most possible movement sounds are not pleasing. What good composers know in their bones is which movement sounds are expressive, and which sequences of movement sounds tell an evocative story. But they also know even deeper in their bones which sounds sound like humans moving, and *that* is what we'll be discussing next, in the upcoming chapter and in the Encore.

# CHAPTER 4

# Musical Movement

## SERIOUS MUSIC

This dictionary of musical themes, by Harold Barlow and Sam Morgenstern, supplies an aid which students of music have long needed. . . . We should now have something in musical literature to parallel Bartlett's *Familiar Quotations*. Whenever a musical theme haunted us, but refused to identify itself no matter how much we scraped our memory, all we should have to do would be to look up the tune in Barlow and Morgenstern, where those ingenious dictionary-makers would assemble some ten thousand musical themes, with a notation-index or theme-finder, to locate the name of the composition from which the haunting fragment came, and the name of the composer.

– John Erskine, 1948, in the preface to
Barlow and Morgenstern's *A Dictionary of Musical Themes*.

In the 1940s it must have been laborious to construct a dictionary of musical themes, but that's what Barlow and Morgenstern went ahead and did. It is unclear whether anyone ever actually used it to identify the tunes that were haunting them, and, at any rate it is obsolete today, given that our iPhones can tell us the name and composer of a song if you merely sing a few bars. The iPhone software is called "Shazam", a great advance over locutions such as, "Hey, can you BarlowandMorgenstern this song for me?" Now, in defense of Barlow and Morgenstern, Shazam does not recognize much classical music, which makes me the life of the party when someone's Shazam comes up empty-handed in the attempt to identify what the pianist is playing, and I pull out my 642-page Barlow and Morgenstern and tell them it is Chopin's Concerto No. 1 in E Minor. And, I add, it is the third theme occurring within the second movement . . . because that's how I roll.

The other great use I have found for Barlow and Morgenstern's dictionary is as a test bed for the movement theory of music. Each of its 10,000 themes nicely encapsulates the fundamental part of a tune—no chords, no harmony, no flourishes. Most themes have around one to two dozen notes, and so, in movement terms, they correspond to short bouts of behavior. (Figure 18 shows three examples of themes from Barlow and Morgenstern.) There are at least two good reasons for concentrating my efforts on this data set.

**FIGURE 18.** Example themes from the Barlow & Morgenstern dictionary. *Top*: A theme from Bach's Partita, No. 1 in B minor. *Middle*: A theme from Beethoven's Sonata No. 7 in D. *Bottom*: A theme from Sibelius's Quartet Op. 56 "Voces Intimae".

First, the dictionary possesses a *lot* of themes—10,000 of them. This is crucial for our purposes because we're studying messy music, not clean physics. One can often get good estimates of physical regularities from a small number of measurements, but even though (according to the music-is-movement theory) music's structure has the signature of the physical regularities of human movement, music is one giant leap away from physics. Music is the product of cultural selection among billions of people, thousands of years, and hundreds of cultures, and so we can only expect to see a blurry signature of human movement inside any given piece or genre of music. On top of that, we have the wayward ways of composers, who are often bent on marching to their own drum and *not* fitting any pattern they might notice in the works of others. Music thus is inherently even messier than speech, and that's why we need a lot of tunes for our data. With enough tunes, we'll be able to see the moving humans through the fog.

The *Dictionary of Musical Themes* is also perfect for our purposes here because it is a dictionary of *classical* music. "What's so great about classical music?" you might ask. Nothing, is the answer. Or, at least, there is nothing about the category of classical music that makes it more worthy of study than other categories of music. But it is nevertheless perfect for our purposes, and for an "evolutionary" reason. We are interested in analyzing not just any old tune someone can dream up, but the tunes that actually get selected for. We want our data set to have the "melodic animals" that have fared well in the ecology of minds they inhabit. Classical music is great for this because it has existed as a category of music for several centuries. The classical music that survives to be played today is just a tiny fraction of all the compositions written over the centuries, with most composers long dead—and even longer obscure.

Ultimately, the theory developed here will have to be tested on the broad spectrum of music styles found across humankind, but, for the reasons I just mentioned, Western classical

music is a natural place to begin. And who is going to be motivated to analyze broad swaths of music for signs of human
movement if their curiosity is not at least piqued by the success
of the theory on a data set closer to home? As it happens, for
many of the analyses carried out in the following chapters, we
did also analyze a database of approximately 10,000 Finnish
folk songs. The results were always qualitatively the same,
and I won't discuss them much here. At any rate, Finnish folk
are universally agreed to be a strange and taciturn people,
and they are (if just barely) in the West, so they don't really
broaden the range of our musical data.

With the Barlow&Morgenstern app installed in our toolkit,
and with good Finyards slandered without reason, we are ready
to embark on a quest for the signature of expressive human
movers in music.

In this chapter we will successively take on rhythm, pitch,
and loudness. As we will see, when we humans move, we have
our own signature rhythm, pitch modulations, and loudness
fluctuations. I will introduce these fingerprints of human
movement, and provide evidence that music has the same
fingerprints. I have at this point accumulated more evidence
than can be reasonably included in this chapter, and so I have
added an "Encore" chapter at the end of the book which takes
up many other converging lines of evidence for human movement hidden inside music.


## DRUM CORE

When most people think about the auditory features peculiar
to music, they are likely to focus on melody, and in particular
upon the melodic contours, or the pattern of pitch rises and
falls. Perhaps this bias toward melody is because the most
salient visible feature of written music is that the notes go up
and down on the staff. Or maybe it is because our fingers go

up and down our instruments, pressing different buttons for different pitches; or because much of the difficulty in playing an instrument is learning to move quickly from pitch to pitch. Whatever the reason, the pitch modulations of the melody get a perceived prominence in music. This is an eternal thorn in the side of percussionists, often charged with not *really* playing an instrument, and of rappers, dismissed as not *really* being musicians.

But in reality, the chief feature of music is not the pitch contours of melody at all, but rhythm and beat, which concern the timing, emphasis, and duration of the notes. Whereas nearly all music has a rhythm and a beat, music can get by without melodic pitch modulations. I just came back from a street fair, for example, where I heard a rock band, an acoustic guitarist, and a drum group. All three had a rhythm and beat, but only two of the three had a melody. The drum group had no melody, but its rhythm and beat made it music—the best music at the fair, in fact.

The rhythm-and-beat property is the hard nugget at the core of music. And the diamond at the very center of that nugget is the beat, all by itself. Let's begin our examination of musical structure, then, with the beat.

We humans make a variety of beat-like sounds, including heartbeats, sexual gyrations, breathing, and certain vocalizations like laughing and sobbing. But one of the most salient beat-like sounds we make is when we walk, and our feet hit the ground over and over again in a regular repeating pattern. *Hit-ring, hit-ring, hit-ring,* or *boom, boom, boom.* Such beat-like gaits resounding from a mover are among the most important sound patterns in our lives, because they are the centerpiece of the auditory signature of a human in our vicinity, maybe a potential lover, murderer, or mailman. This is why the beat is so fundamental to music: natural human movement has a beat, and so music must have a beat. That is, from the music-is-movement theory's point of view, a beat must be as integral to music as footstep sounds are to human movement. And because most actions we carry out have regularly repeating footsteps, most music will have a beat.

And music is not merely expected to have a regularly repeating beat, but to have a *human* step-like beat. Consider the following three *prima facie* similarities between musical beat and footsteps. First, note that the rate of musical beats tends to be around 1 to 2 beats per second, consistent with human footstep rates. Second, also like human footsteps, the beat need not be metronome-like in its regularity; rather, the beat can have irregularities and still be heard as a beat, because our auditory footstep-recognition mechanisms don't expect perfectly metronome-like human movers. In fact, musical performers are known to sometimes purposely *add* irregularities to the beat's timing, with the idea that it sounds better. And a third initial similarity between footsteps and musical beats is that when people go from moving to not moving, the rate of their footsteps slows down, consistent with the tendency toward a slowing of the beat (*ritardando*) at the end of pieces of music (a topic of study by researchers such as Henkjan Honing, Jacob Feldman, and others over the years). Not all objects stop in this fashion: recall from Chapter 2, on solid-object physical events, that a dropped ball bounces with ever *greater* frequency as it comes to a stop. If musical beat were trying to mimic simple solid object-sounds instead of human movers, then musical endings would undergo *accelerando* rather than *ritardando*. But that's not how humans slow down, and it's not how music slows down.

In addition to beats being footstep-like in their rate, regularity, and deceleration, beats are footstep-like in the way they are *danced* to. Remember those babies shaking their stinky bottoms that we discussed in the previous chapter's section titled "Motionally Moving"? They dance, indeed, but one might suspect that they aren't very good at it. After all, these are babies who can barely walk. But baby dancers are better than you may have realized. While they're missing out on the moves that make *me* a sensation at office parties, they get a lot right. To illustrate how good babies are at dancing, consider one fundamental thing you do *not* have to tell them: dance *to* the beat. Babies gyrate so that their body weight tends to be lower to the ground on, and only on, every beat. They somehow "realize" that to dance means not merely to be time-locked to the music, but to give special footstep status to the beat. Babies don't, for example, bounce to every other beat, nor do they bounce twice per beat. And dancing to the beat is something we adults do without ever realizing that there are other possibilities. MCs never yell, "Three steps to the beat!" or "Step in between the beat!" or "Step on the sixteenth note just after the beat, and then again on the subsequent thirty-second note!" Instead, MCs shout out what every toddler implicitly knows: "Dance *to* the beat!" The very fact that we step to the beat, rather than stepping in the many other time-locked ways we could, is itself a fundamental observation about the relationship between movement and music, one that is difficult to notice because of the almost tautological ring to the phrase, "Step to the beat." Why we tend to step to the beat should now be obvious, given our earlier discussion about the footstep-like meaning of the beat, and (in the previous chapter) about dance music sounding like contagious expressive human behaviors. We step to the beat because our brain thinks we are matching the gait of a human mover in our midst.

Recall the drum group at the festival I mentioned near the start of this section. There is something I didn't mention: there *was* no group. More exactly, there was a tent exhibition with a large variety of percussion instruments, and the players were children and adults who, upon seeing and hearing the drums, joined in the spontaneous jam sessions. These random passers-by were able to, and wanted to, make rhythms matching those around them. Watching this spectacle, it almost seems as if we humans are born to drum. But is it so surprising that we're able to drum to the beat if our actions are the origins of the very notion of the beat?

Before discussing further similarities between beats and footsteps, we need to ask about all the notes occurring in music that are *not* on the beat. Beat may be fundamental to music, but I doubt I'd be bothering to write this—or you to read it—if music were always a simple, boring, one-note-per-beat affair. It is the total pattern of on-beat and off-beat notes that determines a piece of music's rhythm, and we must address the question: if on-beat notes are footsteps, then what human-movement-related sounds might the off-beat notes sound like?

## GANGLY NOTES

The repetitive nature of our footsteps is the most fundamental regularity found in our gait, explaining the fundamental status of the beat in music. But we humans make a greater racket than we are typically consciously aware of. Much of our body weight consists of four heavy, gangly parts—our limbs—and when we are on the move, these ganglies are rattling about, bumping into all sorts of things. When our feet swing forward in a stride, they float barely above the ground, and very often shuffle on their way to landing. In natural terrain, the grass, rocks, dirt, and leaves can get smacked or brushed in between the beat. Sometimes one's own body hits itself—legs hitting each other as they pass, or arms hitting the body as they swing. And often we are carrying things, like a quiver of arrows, a spear, a keychain, or a sack of wet scalps of the neighboring villagers, and these will clatter and splat about as we move.

Not only do our clattering ganglies clang in between our footsteps, they make their sounds in a time-locked fashion to the footsteps. This is because when we take a step, we *initiate* a "launch" of our limbs (and any other objects carried on our bodies) into a behavior-specific "orbit," an orbit that will be repeated on the next step if the same behavior is repeated. In some cases the footstep causes the gangly hit outright, as when our step launches our backpack a bit into the air and it then thuds onto our back. But in other cases the step doesn't directly cause the between-the-beat gangly hit so much as it triggers a sequence of motor events, such as our arms brushing against our body, which will recur at the same time delay after the next step. Exactly what the time delay will be after the step depends on the specific manner in which any given gangly part (appendage, carried object, or carried appendage) swings and bounces, which in turn depends on its physical dimensions, how it hangs, where on the body it lies, and how it participates in the behavior.

From the auditory pattern of these footstep–time-locked clattering ganglies, we are able to discern what people are doing. Walking, jogging, and running sound different in their patterns of hits. A sharp turn sounds different from a mover going straight. Jumping leads to a different pattern, as does skipping or trotting. Going up the stairs sounds distinct from going down. Sidestepping and backing up sound different than forward movement. Happy, angry and sad gaits sound different. Even the special case we discussed in the previous chapter—sex—has its own banging ganglies. Close your eyes while watching a basketball game on television, and you'll easily be able to distinguish times when the players are crossing the court from times when they are clustered on one team's side; and you will often be able to make a good guess as to what kind of behavior, more specifically, is being displayed at any time. You can distinguish between the pattern of hits made by a locomoting dog versus cat, cow versus horse. And you can tell via audition whether your dog is walking, pawing, or merely scratching himself. It should come as no surprise that you have fine-grained discrimination capabilities for sensing with your ears the varieties of movements we *humans* make, movements we hear in the pattern of gangly bangings.

If the pattern of our clanging limbs is the cue our auditory system uses to discern a person's type of behavior, then music that has culturally evolved to sound like human movement should have gangly-banging-like sounds in it. And just as gangly bangings are time-locked to the steps, music's analog of these should be time-locked to the beat. And, furthermore, musical banging ganglies should be crucial to the identity of a song, just as the pattern of a mover's banging ganglies is crucial to identifying the type of behavior.

Where are these banging ganglies in music? Right in front of our ears! Musical banging ganglies are simply *notes*. The notes on the beat sound like footsteps (and are typically given greater emphasis, just as footsteps are more energetic than between-the-steps body hits), and the notes occurring between the beats are like the other body hits characterizing a mover's behavior. Beats are footsteps, and rhythm (more generally) is the pattern of a mover's banging ganglies. Just as between-the-steps body-hit sounds are time-locked to footsteps, notes are time-locked to the beat. And, also like our gait, pieces of music that have the same sequence of pitches but differ considerably in rhythm are perceived to be different tunes. If we randomly change the note durations found in "Twinkle, Twinkle Little Star," thereby obliterating the original rhythm, it will no longer *be* "Twinkle, Twinkle Little Star." Similarly, if we randomly change the timing of the pattern of banging ganglies for a basketball player going up for a layup, it will no longer *be* the sound of a layup.

Rhythm and beat have, then, some similarities to the structure of our banging ganglies. We will discuss more similarities in the upcoming sections and in the Encore chapter. But there is one important similarity that might appear to be missing: musical notes usually come with a *pitch*, and yet our footsteps and gangly hits are not particularly pitchy. How can the dull thuds of our bodies possibly be pitchy enough to explain the central role of pitch in music?

If you have already read the earlier chapter on speech, then you may have begun to have an appreciation for the rings occurring when any solid-object physical event occurs. As we discussed, we are typically not consciously aware of the rings, but our auditory system hears them and utilizes them to determine the identity of the objects involved in events (e.g., to tell the difference between a pencil and a paper clip hitting a desk). Although the pitch of a typical solid object may not be particularly salient, it can become much more salient when contrasted with the distinct pitches of other objects' rings. For example, a single drum in a set of drums doesn't sound pitchy, but when played in combination with larger and smaller drumheads, each drum's pitch becomes easy to hear. The same is true for percussionists who use everyday objects for their drums—in such performances one is always surprised to hear the wide range of pitches occurring among all the usually pitchless-seeming everyday objects. Our footsteps and banging ganglies *do* have pitches, consistent with the hypothesis that they are the fundamental source of musical notes. (As we will see, these gangly pitches are analogous to chords, not to melody—which, I will argue later, is driven by the Doppler effect.)

If I am right that musical notes have their origin in the sounds that humans make when moving, then notes should come in human-gait-like *patterns*. In the next section, we'll take up a simple question in this regard: does the number of notes found between the beats of music match the number of gangly bangs between footsteps?

## THE LENGTH OF YOUR GANGLY

Every 17 years, cicadas emerge in droves out of the ground in Virginia, where I grew up. They climb the nearest tree, molt, and emerge looking a bit like a winged tank, big enough to fill your palm. Since they're barely able to fly, we used to set them on our shoulders on the way to school, and they'd often not bother to fly away before we got there. And if they did fly, it wasn't *really* flying at all. More of an extended hop, with an exoskeleton-shaking, tumble-prone landing. With only a few days to live, and with billions of others of their kind having emerged at the same time, all of them screeching mind-numbingly away, they didn't need to go far to find a mate, and graceful flight did not seem to be something the females rewarded.

Cicadas have, then, a distinctively cicada-like sound when they move: a leap, a clunky clatter of wings, and a heavy landing (often with further hits and skids afterward). The closest thing to a footstep in this kind of movement is the landing thud, and thus the cicada manages to fit dozens of banging ganglies—its wings flapping—in between its land-ings. If cicadas were someday to develop culture and invent music that tapped into their auditory movement-recognition mechanisms, then their music might have dozens of notes between each beat. With *Boooom* as their beat and *da* as their wing-flap inter-beat note, their music might be something like "*Boooom-da-da-da-da-da-da-da-da-da-da-da-da-da-da-da-da-da-da-da-da-da-da-da-da-da-da-da-da-da-da-da-da-da-Boooom-da-da-da-da-da-da-da-da-da-da-da-da-da-da-da-da-da-da-da-da-da-da-da-da-da-da-da-da-da*," and so on. Perhaps their ear-shattering, incessant mating call *is* this sound!

Whereas cicadas liberally dole out notes in between the beats, Frankenstein's monster in the movies is a miser with his banging ganglies, walking so stiffly that his only gait sounds are his footsteps. Zombies, too, tend to be low on the scale of banging-gangly complexity (although high on their intake of basal ganglia).

When *we* walk, our ganglies are more complex than those of Frankenstein and his zombie dance buddies, but ours are doled out much more sparingly than the cicadas'. During a step, your leg swings forward just once, and so it can typically only get one really good bang on something. More complex behaviors can lead to more bangs per step, but most commonly, our movements have just one between-the-footsteps bang—or none. Our movements tend to sound more like the following, where "*Boooom*" is the regularly repeating footstep sound and "da" is the between-the-steps sound: "*Boooom-Boooom-Boooom-da-Boooom-Boooom-da-Boooom-da-Boooom-da-da-Boooom-da-Boooom-da-Boooom.*" (Remember to do the "*Boooom*" on the beat, and cram the "*da*"s in between the beats.)

Given our human tendency to make roughly zero to one gangly bang between our steps, our human music should tend to pack notes similarly lightly between the beats. Music is thus predicted to tend to have around zero to one between-the-beats note. To test for this, we can look at the distribution of time gaps between musical notes. If music most commonly has about zero to one note between the beats—along with notes usually on the beat—then the most common note-to-note time gap should be in the range of a half beat to a beat.

To test this, as an RPI graduate student Sean Barnett analyzed an electronic database of Barlow and Morgenstern's 10,000 classical themes, the ones we mentioned at the start of this chapter. For every adjacent pair of notes in the database, Sean recorded the duration between their onsets (i.e., the time from the start of the first note to the start of the second note). Figure 6 shows the distribution of note-to-note time gaps in this database—which time intervals occur most commonly, and which are more rare. The peak occurs at ½ on the x axis, meaning that the most common time gap is a half-beat in length (an eighth note). In other words, there is one note between the beats on average, which is broadly consistent with expectation.

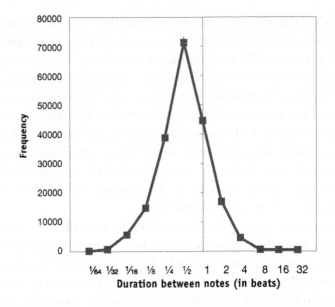

**FIGURE 19.** The distribution of durations between notes (measured in beats), for the roughly 10,000 classical themes. One can see that the most common time-gap between notes is a half beat long, meaning on average about one between-the-beat note. This is similar to human gait, typically having around zero to one between-the-step "gangly" body-hit.

We see, then, that music tends to have the number of notes per beat one would expect if notes are the sounds of the gan-glies of a human—not a cicada, not a Frankenzombie—mover. Musical notes are gangly hits. And the beat is that special gangly hit called the footstep. In the next section we will dis-cuss some of what makes the beat special, and see if footsteps are similarly special (relative to other kinds of gangly hits).

## BACKBONE

My family and I just moved into a new house. Knowing that my wife was unhappy with the carpet in the family room, and knowing how much she fancies tiled floor, I took the day off and prepared a surprise for her. I cut tile-size squares from the carpet, so that what remained was a checkerboard pattern, with hardwood floors as the black squares and carpet as the white squares.

I couldn't sleep very well that night on the couch, and so I headed into the kitchen for a bite. As I pondered how my plan had gone so horribly wrong, I began to notice the sounds of my gait. Walking on my newly checkered floor, my heels occasionally banged loudly on hard wood, and other times landed silently on soft carpet. Although some of my between-step intervals were silent, between many of my steps was a strong bump or shuffle sound when my foot banged into the edge of the two-inch-raised carpet. The overall pattern of my sounds made it clear when my footsteps *must* be occurring, even when they weren't audible.

Luckily for my wife—and even more so for me—I never actually checkered my living room carpet. But our world is itself checkered: it is filled with terrain of varying hardness, so that footstep loudness can vary considerably as a mover moves. In addition to soft terrain, another potential source of a silent step is the modulation of a mover's step, perhaps purposely stepping lightly in order to not sprain an ankle on a crooked spot of ground, or perhaps adapting to the demands of a particular behavioral movement. Given the importance of human footstep sounds, we should expect that our auditory systems were selected to possess mechanisms capable of recognizing human gait sounds even when some footsteps are missing, and to "fill in" where the missing footsteps are, so that the footsteps are perceptually "felt" even if they are not heard.

If our auditory system can handle missed footsteps, then we should expect music—if it is "about" human movement—to tap into this ability with some frequency. Music should be able to "tell stories" of human movement in which some footsteps are inaudible, and be confident that the brain can handle it. Does music ever skip a beat? That is, does music ever *not* put a note on a beat?

Of course. The simplest cases occur when a sequence of notes on the beat suddenly fails to continue at the next beat. This happens, for example, in "Row, Row, Row Your Boat," when each "row" is on the beat, and then the beat just after "stream" does not get a note. But music is happy to skip beats in more complex ways. For example, in a rhythm like that shown in Figure 20, the first beat gets a note, but all the subsequent beats do not. In spite of the fact that only the first beat gets a note, you *feel* the beat occurring on all the subsequent skipped beats. Or the subsequent notes may be perceived to be *off*-beat notes, not notes on the beat. Music skips beats and humans miss footsteps—and in each case our auditory system is able to perceptually insert the missing beat or footstep where it belongs. That's what we expect from music if beats are footsteps.

FIGURE 20. The first note is on the beat, but because it is an eighth note (lasting only half a beat), all the subsequent quarter notes (which are a beat in length) are struck on the off beat. You feel the beat occurring between each subsequent note, despite there being no note on the beat.

The beat is the solid backbone of music, so strong it makes itself felt even when not heard. And the beat is special in other ways. To illustrate this, let's suppose you hear something strange approaching in the park. What you find unusual about the sound of the thing approaching is that each step is quickly followed by some other sound, with a long gap before the next step. *Step-bang . . . . . . Step-bang . . . . . .* "What on Earth *is* that?" you wonder. Maybe someone limping? Someone walking with a stick? Is it human at all?! The strange mover is about to emerge on the path from behind the bushes, and you look up to see. To your surprise, it is simply a lady out for a stroll. How could you not have recognized that?

You then notice that she has a lilting gait in which her forward-swinging foot strikes the ground before rising briefly once again for its proper footstep landing. She makes a hit sound immediately *before* her footstep, not immediately after as you had incorrectly interpreted. *Step* . . . . . . *bang-Step* . . . . . . *bang-Step* . . . . . . Her gait does indeed, then, have a pair of hit sounds occurring close together in time, but your brain had mistakenly judged the first of the pair of sounds to be the footstep, when in reality the *second* in the pair was the footstep. The first was a mere shuffle-like floor-strike during a leg stride. Once your brain got its interpretation off-kilter, the perceptual result was utterly different: lilting lady became mysterious monster.

The moral of this lilting-lady story is that to make sense of the gait sounds from a human mover, it is *not* enough to know the temporal pattern of gait-related hit sounds. The lilting lady and mysterious monster have the *same* temporal pattern, and yet they sound very different. What differs is *which* hits within the pattern are deemed to be the *footstep* sounds. Footsteps are the backbone of the gait pattern; they are the pillars holding up and giving structure to the other banging gangly sounds. If you keep the temporal pattern of body hits but shift the backbone, it means something very different about the mover's gait (and possibly about the mover's identity). And this meaning is reflected in our perception.

If musical rhythm is like gait, then the feel of a song's rhythm should depend not merely on the temporal pattern of notes, but also on where the beat is within the pattern. This is, in fact, a well-known feature of music. For example, consider the pattern of notes in Figure 21.

FIGURE 21. An endlessly repeating rhythm of long, short, long, short, etc., but with neither "long" nor "short" indicated as being on the beat. One *might* have thought that such a pattern should have a unique perceptual feel. But as we will see in the following figure, the pattern's feel depends on where the beat-backbone is placed onto it. Human gait is also like this.

One might think that such a never-ending sequence of long-short note pairs should have a single perceptual feel to it. But that same pattern sounds very different in the two cases shown in Figure 9, which differ only in whether the short or the long note marks the beat. The first of these sounds jarring and inelegant compared to the second. The first of these is, in fact, like the mysterious monster we imagined approaching a moment ago, and the second is like the lilting lady the mover turned out to be.

FIGURE 22. (a) A short-long rhythm, which sounds very different from (and less natural than) the long-short rhythm in (b).

Music, like human gait-related sounds, cannot have its beat shifted willy-nilly. The identity of a gait depends on *which* hits are the footsteps, and, accordingly, the identity of a song depends on which notes are on the beat. And when a beat is not heard, the brain infers its presence, something the brain also does when a mover's footstep is inaudible.

There are, then, a variety of suspicious similarities between human gait and the properties of musical rhythm. In the upcoming section, we begin to move beyond rhythm toward melody and pitch. We'll get there by way of discussing how chords may fit within this movement framework, and how choreography depends on more than just the rhythm.

Although we're moving on from rhythm now, there are further lines of evidence that I have included in the Encore, which I will only provide teasers for here:

**Encore 1: "The Long and Short of Hit"** Earlier in this section I mentioned that the short-long rhythm of the mysterious monster sounds less natural than the long-short rhythm of the lilting lady. In this part of the Encore, I will explain why this might be the case.

**Encore 2: "Measure of What?"** I will discuss why changing the measure, or time signature, in music modulates our perception of music.

**Encore 3: "Fancy Footwork"** When people change direction while on the move, their gait often can become more complex. I show that the same thing occurs in music: when pitch changes (indicative, as we will see, of a *turning* mover), rhythmic complexity rises.

**Encore 4: "Distant Beat"** The nearer movers are, the more of their gait sounds are audible. I will discuss how this is also found in music: louder portions of music tend to have more notes per beat.


## GANGLY CHORDS

Earlier in this chapter, we discussed how footsteps and gangly bangs ring, and how these rings tend to have pitches. I hinted then that it is the Doppler shifting of these pitches which is the source of melody, something we will get to soon in this chapter. But we have yet to talk about the other principal role of pitch in music—harmony and chords.

When pitches combine in close temporal proximity, the result is a distinct kind of musical sound called the chord. For example, C, E, and G pitches combine to make the C major chord. Where do chords fit within the music-is-movement theory? To begin to see what aspect of human movement chords might echo, consider what happens when a pianist wants to get a rhythm going. He or she could just start tapping the rhythm on the wood of the piano top, but what the pianist actually does is play the rhythm via the piano keys. The rhythm is implemented with pitches. And furthermore, the pianist doesn't just bang out the rhythm with any old pitches. Instead, the pianist picks a *chord* in which to establish the rhythm and beat. What the pianist is doing is analogous to what a guitarist does with a strum. Strums, whether on a guitar or a piano, are both rhythm *and* chord.

My suspicion is that rhythm and chords are two distinct kinds of information that come from the gangly banging sounds of human movers. I have suggested in this chapter that rhythm comes from the temporal pattern of human banging ganglies. And now I am suggesting that chords come from the combinations (or perhaps the constituents) of pitches that occur among the banging gangly rings. Gait sounds have temporal patterns *and* pitch patterns, and these underlie rhythm and chords, respectively. And these two auditory facets of gait are informative in different ways, but both broadly within the realm of "attitude" or "mood" or "intention," as opposed to being informative about the direction or distance of the mover—topics that will come up later in regard to melody and loudness, respectively.

If rhythm and chords are each aspects of the sounds of our ganglies, then we should expect chords to cycle through their pitches on a time scale similar to that of the rhythm, and time-locked to the rhythm; the rhythm and chord should have the same time signature. For example, in an Alberti chord/rhythm pattern, one's left hand on the piano might play the notes [**C**GEG][**C**GEG][**C**GEG], where each set of square brackets shows a two-beat interval, and bold type and underlines indicate the emphases in the rhythm. One can see that the same two-beat pitch pattern and rhythm repeats over and over again. The pitch sequence and the rhythm have the same

²/₄ time signature. It is much rarer to find chords expressed in a way that mismatches the rhythm, such as the following case, where the chord is expressed as a repeated pattern of three pitches—C-G-E—and thus the two-beat rhythm cycles look like [**C**GEC] [**G**ECG] [**E**CGE]. In this case, notice that the first two-beat interval—the rhythm's cycle—has the pitch sequence CGEC, but that the second one has, instead, GECG. The pitch cycle for the chord is not matched to the rhythm's cycle. In real music, if the rhythm is in ²/₄ time, then the chord will typically not express itself in ¾ time. Rhythm and chords tend to be locked together in a way that suggests they are coming from the same worldly source, and therefore the arguments in this chapter lead one to speculate that both rhythm *and* chords come from, or are about, our gangly banging sounds.

We can also ask which pitch within the expressed chord is most likely to be the one played on the beat. For human movers, the lowest-pitched gangly bang we make is usually our footsteps. For music and the rhythmic expression of chords, then, we expect that the pitch played on the beat will tend to be lower than that played between the beats. Indeed, chords are usually caressed starting on the lowest expressed pitch (and often on the chord's tonic, which in a C major chord would be the C pitch). Chords are, again, like gangly rings, with the lowest pitch ringing on the beat.

Consider yet another attribute of human gait: our gangly bangings can occur *simultaneously*. Multiple parts of a mover's body can be clattering at the *same* time, and even a single bang will cause a ring on both the banger and the banged. So we should expect that the auditory mechanisms evolved for sensing gait would be able to process gait from the input of multiple simultaneous pitches. Consistent with this, the pitches within a chord are commonly played simultaneously, and our brains can make perfect sense of the simultaneously-occurring notes. Pitch modulations that are part of the melody, on the other hand, almost never occur simultaneously (as we will discuss later).

The idea that musical chords have their foundation in the pitch combinations heard in the banging gangly sounds of human movers is worth investigating further. However, there are a wide variety of phenomena concerning chords that one would hope to explain, and that I currently have no theoretical insights into how to explain based on the raw materials of our ganglies. The laboratory of Dale Purves at Duke University has carried out exciting research suggesting that the human voice may explain the signature properties of the diatonic scale, and one might imagine persuasive explanations for chords emerging from his work. In fact, people do often vocalize while they move and carry out behaviors, and one possibility is that chords are not about gangly bangs at all, but about the quality of our vocalizations. The advantage of looking to gangly bangs as the foundation for chords, however, is that banging ganglies are time-locked to footsteps, and thus intrinsically note-like. Human vocalizations, however, are not time-locked to our footsteps, and also lack a clear connection to the between-steps movements of our banging ganglies. If chords were driven by vocalizations, we would not be able to explain why chords are so wedded to the rhythm, as demonstrated above. If one can find chords in our ganglies, then it allows for a unified account: our banging ganglies would explain *both* rhythm and chords—and the tight fit between them.

Chords, I have suggested, may have their origins in the pitches of the complex rings given off by gangly human movers. Later in the chapter, I will suggest that the pitch modulations in melody, in contrast, come from the Doppler shifting of the envelope of those gangly pitches.


## CHOREOGRAPHED FOR YOU

Choreography is all about finding the right match between human movement and music. I had always figured it wasn't the music as a whole that must match people's movement so much as it was just the *rhythm and beat.* Get the music's bangs in line with the people's bangs—that's all choreographers needed to care about. But I now realize there's a great deal more to it. A lot of what matters in good choreography is not the rhythm and beat at all. The melodic contour matters too, and so does the loudness. (To all you choreographers who already know this, please bear with me!)

Why should musical qualities beyond rhythm and beat matter to choreography? Because there are sound qualities beyond our intrinsic banging gangly sounds that also matter for sensing human movers. For example, suppose you and I are waiting for an approaching train, but you are 100 yards farther up the tracks (toward the approaching train) than I am. You and I will hear the same train "gait" sounds—the chugs, the rhythmic clattering of steel, and so on—but *you* will hear the train's pitch fall (due to the Doppler effect) before I hear it fall. Now imagine that I am wearing headphones connected to a microphone on your lapel, so that at my position along the tracks I am listening to the sounds *you* are hearing at your position along the tracks. The intrinsic gait sounds of the train would be choreographed appropriately with my visual perception of the train, because those gait sounds don't depend on the location of the listener. But my headphone experience of the *pitch and loudness* contours would no longer fit my visual experience. The train's pitch now begins falling too early, and will already approach its lowest going-away-from-me pitch before the train even reaches me. The train's loudness is also now incorrect, reaching its peak when the train is still 100 yards from reaching me. This would be a deeply ecologically incoherent audiovisual experience; the auditory stream from the headphones would not be choreographed with the train's visible movements, even though the temporal properties—the beat and rhythm—of the train's trangly bangings are just as they should be.

Real-world choreography pays attention to pitch and loudness contours as well as gait sounds; and, crucially, which pitch and loudness contour matches a movement depends on where the listener is. Choreography for pitch and loudness contours is listener-centric.

The implication for musical choreography is this: in matching music to movement, the choreographer must make sure that the *view*point is the same point in space as the *listening* point. Good choreography must not merely "know its audience," but know *where* they are. Music choreographed for *you*, where you're sitting, may not be music choreographed for *me*, where I'm sitting. In television, choreographers play to the *camera's* position. If movers are seen in a video to veer toward the camera, melody's pitch must rise to fit the video, for example. In live shows, choreographers play to the audience, although this gets increasingly difficult the more widely the audience is distributed around the stage. (This is one of many reasons why most Super Bowl halftime shows suck.)

Whereas our discussion so far has concerned rhythm and beat, which do not depend on the listener's position, the upcoming sections concern pitch and loudness, each of which depends crucially on the location of the listener. Music with only a beat and a rhythm is a story of human behavior, but without any particular viewpoint. In contrast, music with pitch and loudness modulations puts the listener at a fixed viewpoint (or listening point) in the story, as the fictional mover changes direction and proximity to the listener. These are the mover's kinematics, and the rest of this chapter examines how music tells stories about the kinematics.

## MOTORCYCLE MUSIC

Next time you're on the highway at 70 mph next to a roaring Harley, roll down your window and listen (but do not breathe!).

I did this just the other day, and was struck by something strange about how the chopper sounded. The motorcycle's "footsteps" were there, namely the sounds made by the bike's impacts directly on the asphalt as it barreled over crevices, crags, and cracks. The motorcycle's "banging gangly" sounds were also present—the sounds made by the bike's parts interacting with one another, be they moving parts in the engine or body parts rattling due to engine or road vibrations. And the bike's exhaust pipe also made its high-frequency *vroom* (not quite analogous to a sound made by human movers). These motorcycle sounds I heard were characterized not only by their rhythm, but also by the suite of pitches among the rings of these physical interactions: the bike's "chords." These rhythm and chord sounds informed me of the motorcycle's "state": it is a motorcycle; it is a Harley; it is going over uneven ground; it is powerful and rugged; it needs a bath; and so on. Rhythm and beat (and the chords with which they seem inextricably linked), the topics of much of this chapter thus far, are all about the *state* of the mover—the nature of the mover's gait, and the emotion or attitude expressed by that manner of moving about.

What, then, was so strange about the motorcycle sounds I heard while driving alongside? It was that the motorcycle's overall pitch and loudness were *constant*. In most of my experiences with motorcycles, their pitch and loudness vary dynamically. This is because motorcycles are typically *moving* relative to me (I never ride them myself), and consequently they are undergoing changes in pitch due to the Doppler effect, and changes in loudness due to changing proximity. These pitch and loudness modulations give away the *action*, and *that* was what was missing: the motorcycle had attitude but no action.

Music gets its attitude from the rhythm and beat, but when music wants to tell a story about the mover in *motion*—the mover's kinematics—music breaks out the pitch and becomes melodic, and twiddles with the volume and modulates the loudness. The rest of this chapter is about the ecological origins of melody and loudness. We will begin with melody, but before I begin to defend what I think musical melodic pitch means, we need to overcome a commonly held bias—encoded in the expressions "high" and "low notes"—that musical pitch equates with spatial position.

## WHY PITCH SEEMS SPATIAL

Something is falling from the sky! Quick, what sound is it making? You won't be alone if you feel that the appropriate sound is one with a falling pitch (possibly also with a crescendoing loudness). That's the sound cartoons use to depict objects falling from overhead. But is that the sound falling objects *really* make, or might it be just a myth?

No, it's not a myth. It's true. If a falling object above you is making audible sounds at all (either intrinsically or due to air resistance), then its pitch will be falling as it physically falls for the same reason passing trains have falling pitch: falling objects (unless headed directly toward the top of your head) are *passing* you, and so the Doppler effect takes place, like when a train passes you. Falling objects happen to be passing you in a *vertical* direction rather than along the ground like a train, but that makes no difference to the Doppler effect. Because falling objects have falling pitch, we end up associating greater height with greater pitch. That's why, despite greater sound frequencies not being "higher" in any real sense, it feels natural to call greater frequencies "higher." Pitch and physical height are, then, truly associated with one another in the world.

But the association between pitch and physical height is a *misleading* association, *not* indicative of an underlying natural regularity. To understand why it is misleading, let's now imagine, instead, that an object resting on the ground in front of you suddenly launches *upward* into the sky. How does *its* pitch change? If it really were a natural regularity that higher in the sky associates with higher pitch, then pitch should rise as the object rises. But that is not what happens. The Doppler effect ensures that its pitch actually *falls* as it rises into the sky. To understand why, consider the passing train again, and ask what happens to its pitch once it has already reached its nearest point to you and is beginning to move away. At this point, the train's pitch has already decreased from its maximum, when it was far away and approaching you, to an intermediate value, and it will *continue* to decrease in pitch as it moves away from you. The pitch "falls" or "drops," as we say, because the train is directing itself more and more away from you as it continues straight, and so the waves reaching your ears are more and more spread out in space, and thus lower in frequency. (In the upcoming section, we will discuss the Doppler effect in more detail.) An object leaping upward toward the sky from the ground is, then, in the same situation as the train that has just reached its nearest point to you and is beginning to go away. The pitch therefore *drops* for the upward-launching object. If rocket launches were to be our most common experience with height and pitch, then one would come to associate greater physical height with *lower* pitch, contrary to the association people have now. But because of gravity, objects don't tend to launch upward (at least they didn't for most of our evolutionary history), and so the association between physical height and low pitch doesn't take hold. Objects *do* fall, however (and it is an especially dangerous scenario to boot), and so the association between physical height and "high" pitch wins. Thus, greater height only associates with "higher" pitch because of the gravitational asymmetry; the fundamental reason for the pitch falling as the object falls is the Doppler effect, not physical height at all. Pitch falls for falling objects because the falling object is rushing by the listener, something that occurs also as the train comes close and then passes.

Falling objects are not the only reason we're biased toward a spatial interpretation of pitch (i.e., an interpretation that pitch encodes spatial position or distance). Our music technology—our instruments and musical notation system—accentuates the bias. On most instruments, to change pitch requires changing the position in space of one's hands or fingers, whether horizontally over the keys of a piano, along the neck of a violin, or down the length of a clarinet. And our Western musical notation system codes for pitch using the vertical spatial dimension on the staff—and, consistent with the gravitational asymmetry we just discussed, greater frequencies are higher on the page. The spatial modulations for pitch in instrument design and musical notation are very useful for performing and reading music, but they further bang us over the head with the idea that pitch has a spatial interpretation.

There is yet another reason why people are prone to give a spatial interpretation to melody's "rising and falling" pitch, and that is that melody's pitch tends to change in a continuous manner: it is more likely to move to a nearby pitch than to discontinuously "teleport" to a faraway pitch. This has been recognized since the early twentieth century, and in *Sweet Anticipation* Professor David Huron of Ohio State University summarizes the evidence for it. Isn't this pitch continuity conducive to a spatial interpretation? After all, humans don't jump down hills or fly up them; we walk (or run), step by step. Pitch continuity is at least *consistent* with a spatial interpretation. (But, then again, continuity is consistent with *most* possible physical parameters, including the *direction* of a mover.)

We see, then, that gravity, musical instruments, musical notation, and the pitch continuity of most melody conspire to bias us to interpret musical pitch in a spatial manner (i.e., where pitch represents spatial position or distance). But like any good conspiracy, it gets people believing something *false*. Pitch is not spatial in the natural world. It doesn't indicate distance or measure spatial position. How "high" or "low" a sound is doesn't tell us how near or far away its source is. I will argue that pitch is not spatial in music, either. But then what *is* spatial in music? If music is about movement, it would be bizarre if it didn't have the ability to tell your auditory brain where in space the mover is. As we will see later in this chapter, music *does* have the ability to tell us about spatial location— that's the meaning of *loudness*.

But we're not ready for that yet, for we must still decode the meaning of melodic pitch. My hypothesis is that hiding underneath those false spatial clues lies the true meaning of melodic pitch: the *direction* of the mover (relative to the listener's position). It is that fundamental effect in physics, the Doppler effect, that transforms the directions of a mover into a range of pitches. In order to comprehend musical pitch, and the melody that pitches combine to make, we must learn what the Doppler effect *is*. We take that up next.

## DOPPLER DICTIONARY

In the summer months, our neighborhood is regularly trawled by an ice cream truck, loudly blaring music to announce its arrival. When the kids hear the song, they're up and running, asking for money. My strategy is to stall, suggesting, for example, that the truck only sells dog treats, or that it is that very ice cream truck that took away their older sister whom we never talk about. But soon they're out the door, listening intently for it. "It's through the woods behind the Johnsons'," my daughter yells. "No, it's at the park playground," my son responds. As the ice cream truck navigates the maze of streets, the kids can hear that it is sometimes headed toward them, only to turn at a cross street, and the kids' hearts drop. I try to allay their heartache by telling them they weren't getting ice cream even if the truck *had* come, but then they perk up, hearing it headed this way yet again.

The moral of this story about my forlorn kids is not just how to be a good parent, but how kids can hear the comings and goings of ice cream trucks. There are a variety of cues they could be using for their ice cream–truck sense, but one of the best cues is the truck's *pitch*, the entire envelope of pitches which modulates as it varies in direction relative to my kids' location, due to the Doppler effect.

What exactly *is* the Doppler effect? To understand it, we must begin with a special speed: 768 miles per hour. That's the speed of sound in the Earth's atmosphere, a speed Superman must keep in mind because passing through that speed leads to a sonic boom, something sure to flatten the soufflé he baked for the Christmas party. We, on the other hand, move so slowly that we can carry soufflés with ever so slightly less fear. But even though the speed of sound is not something we need to worry about, it nevertheless has important consequences for our lives. In particular, the speed of sound is crucial for comprehending the Doppler effect, wherein moving objects have different pitches depending on their direction of movement relative to the listener.

Let's imagine a much slower speed of sound: say, two meters per second. Now let's suppose I stand still and clap 10 times in one second. What will you hear (supposing you also are standing still)? You will hear 10 claps in a second, the wave fronts of a 10 Hertz sound. It also helps to think about how the waves from the ten claps are spread out over space. Because I'm pretending that the speed of sound is two meters per second, the first clap's wave has moved two meters by the time the final clap occurs, and so the ten claps are spread out over two meters of space. (See Figure 23a.)

Now suppose that, instead of me standing still, I am moving toward you at one meter per second. That doesn't sound fast, but remember that the speed of sound in this pretend example is two meters per second, so I'm now moving at half the speed of sound! By the time my first clap has gone two meters toward

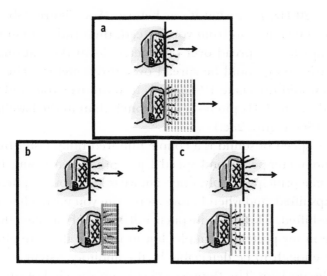

**FIGURE 23.** **(a)** A stationary speaker is shown making ten clap sounds in a second. The top indicates that the wave from the clap has just occurred, not having moved beyond the speaker. In the lower part of the panel, the speaker is in the same location, but one second of time has transpired. The first wave has moved two meters to the right, and the final wave has just left the speaker. A listener on the right will hear a 10 Hz sound. **(b)** Now the speaker is moving in the same direction as the waves, and has moved one meter to the right after one second. The ten claps are thus spread over one meter of space, not two meters as in **(a)**. All ten waves wash over the listener's ears in half a second, or at 20 Hz, twice as in **(a)**. **(c)** In this case the speaker is moving away from the listener, or leftward. By the time the tenth clap occurs, the speaker has moved one meter leftward, and so the ten claps are spread over *three* meters, not two as in **(a)**. Their frequency is thus lower, or 6.7 Hz rather than the 10 Hz in **(a)**.

you, my body and hands have moved one meter toward you, and so my final clap occurs one meter closer to you than my first clap. Whereas my ten claps were spread over two meters when I was stationary, in this moving-toward-you scenario my ten claps are spread over only *one* meter of space. These claps will thus wash over your ears in only *half* a second, rather than a second, and so you will hear a pitch

that is 20 Hz, *twice* what it was before. (See Figure 23b.) If
I were moving *away* from you instead, then rather than my
10 claps being spread over two meters as in the stationary
scenario, they would be spread over three meters. The 10
claps would thus take 1½ seconds to wash over you, and be
heard as a 6.66 Hz pitch—a lower pitch than in the baseline
case. (See Figure 23c.)

The speed of sound is a couple hundred times faster than
the two-meter-per-second speed I just pretended it was, but
the same principles apply: when I move toward you my pitches
are upshifted, and when I move away from you my pitches are
downshifted. The shifts in pitch will be much smaller than
those in my pretend example, but in real life they are often
large enough to be detectable by the auditory system, as we
will discuss later. The Doppler effect is just the kind of strong
ecological universal one expects the auditory system to have
been selected to latch onto, because from it a listener's brain
can infer the direction of motion of a mover, such as an ice
cream truck.

To illustrate the connection between pitch and directedness toward you, let's go back to our generic train example
and assume the track is straight. When a train is far away but
approaching the station platform where you are standing, it
is going almost directly toward you, as illustrated in Figure
24. This is when its pitch will be Doppler shifted upward the
most. (High and constant pitch is, by the way, the signature
of an impending collision.) As the train nears, it gets less
and less directed toward you, eventually to pass you by. Its
pitch thus drops to an intermediate, or baseline, value when
it reaches its nearest point to you and is momentarily moving
neither toward nor away from you (see Figure 24ii). As the
train begins to move away from you, its pitch falls below its
intermediate value and continues to go lower and lower
until it reaches its minimum, when headed directly away (see
Figure 24iii). (If, by the way, you were unwisely standing on

the *tracks* instead of on the platform, then the train's pitch would have remained at its maximum the entire period of time it approached. Then, just after the sound of your body splatting, the train's pitch would instantaneously drop to its lowest pitch. Of course, you would be in no condition to hear this pitch drop.)

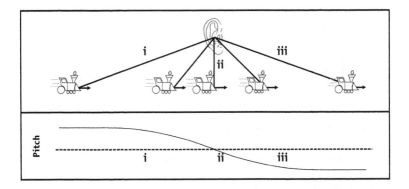

**FIGURE 24.** Illustration that the pitch of a mover (relative to the baseline pitch) indicates the mover's directedness toward you. When the train is headed directly toward the observer, pitch is at its maximum **(i)**, and is at its lowest when headed directly away **(iii)**; in between the pitch is in between **(ii)**.

As a further illustration of the relationship between pitch and mover direction, suppose that a mover is going around in a circle out in front of you (not around you). At (a) in Figure 25 the mover is headed directly away, and so has minimum pitch. The mover begins to turn around for a return, and pitch accordingly rises to a baseline, or intermediate, level at (b). The mover now begins veering toward you, raising the pitch higher, until the mover is headed directly toward you at (c), at which point the pitch is at its maximum. Now the mover begins veering away from you so as not to collide, and pitch falls back to baseline at position (d), only to fall further as the mover moves away to (a) again.

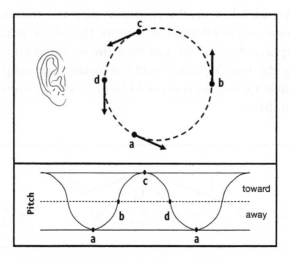

**FIGURE 25.** The upper section shows a mover moving in a circle out in front of the listener (the ear), indicating four specific spots along the path. The lower part of the figure shows the pitch at these four spots on the path. **(a)** When moving directly away, pitch is at its minimum. **(b)** Pitch rises to baseline when at the greatest distance and moving neither toward nor away. **(c)** Pitch rises further to its maximum when headed directly toward the listener. **(d)** Pitch then falls back to baseline when passing tangentially nearby. The pitch then falls back to its minimum again at **(a)**, completing the circle.

From our experience with the train and looping-mover illustrations, we can now build the simple "dictionary" of pitches shown in Figure 26. Given a pitch within a range of pitches, the figure tells us the pitch's meaning: a direction of the mover relative to the listener. In the dictionary of nature, pitch means degree of directedness toward you.

This pitch dictionary is useful, but only to a limited extent. Doppler pitches tend to be fluctuating when you hear them, whether because movers are merely going straight past you (as in Figure 24), or because movers are turning (as in Figure 25). These dynamic pitch changes, in combination with the pitch dictionary, are a source of rich information for a listener. Whereas pitches above and below baseline mean an approaching or receding mover, respectively, *changing* pitch tells us about the mover's turning and veering behavior. A *rising* pitch means that the mover is becoming *increasingly directed* toward the listener; the mover is veering more toward you. And *falling* pitch means that the mover is becoming *decreasingly directed* toward the listener; the mover is veering more away from you. One can see this in Figure 25. From (a) through (c) the mover is veering more toward the listener, and the pitch is rising throughout. In the other portion of the circular path, from (c) to (a) via (d), the mover is veering *away* from the listener, and the pitch is *falling.*

To summarize, pitch informs us of the mover's direction relative to us, and pitch *change* informs us of change of direction—the mover's *veering* behavior. High and low pitches mean an approaching and a receding mover, respectively; rising and falling pitches mean a mover who is veering toward or away from the listener, respectively. We have, then, the following two fundamental pitch-related meanings:

**Pitch**: *Low pitch* means a *receding* mover. *High pitch* means an *approaching* mover.

**Pitch change**: *Falling pitch* means a mover veering more *away*. *Rising pitch* means a mover veering more *toward*.

Because movers can be approaching or receding and *at the same time* veering toward or away, there are 2 × 2 = 4 qualitatively distinct cases, each defining a distinct signature of the mover's behavior, as enumerated below and summarized in Figure 27.

1. Moving away, veering toward.
2. Moving toward, veering toward.
3. Moving toward, veering away.
4. Moving away, veering away.

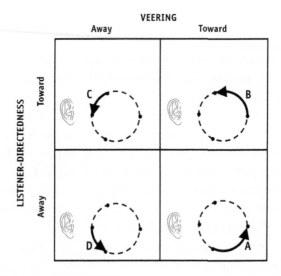

**FIGURE 27.** Four qualitatively distinct categories of movement given that a mover may move toward or away, and may veer toward or away. (I have given them alphabet labels starting at the bottom right and moving counter-clockwise to the other three squares fo the table, although my reason for ordering them in this way won't be apparent until later in the chapter. I will suggest later that the sequence A-B-C-D is a generic, or most common, kind of encounter.

These four directional arcs can be thought of as the fundamental "atoms" of movement out of which more complex trajectories are built. The straight-moving train of Figure 24, for example, can be described as C followed by D, that is, veering away over the entire encounter, but first nearing, followed by receding. (As I will discuss in more detail in the Encore section titled "Newton's First Law of Music," straight-moving movers passing by a listener are effectively veering away from the listener.)

These four fundamental cases of movement have their own pitch signatures, enumerated below and summarized in Figure 28.

(a) **Low, rising pitch** means **moving away, veering toward**.
(b) **High, rising pitch** means **moving toward, veering toward**.
(c) **High, falling pitch** means **moving toward, veering away**.
(d) **Low, falling pitch** means **moving away, veering away**.

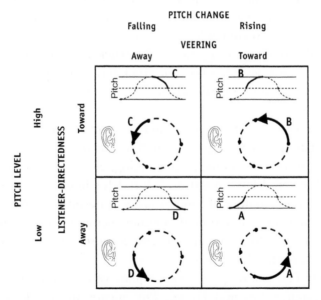

**FIGURE 28.** Summary of the movement meaning of pitch, for low and high pitch, and rising and falling pitch. (Note that I am not claiming people move in circles like shown in the figure. The figure is useful because all movements fall into one of these four categories, which I am illustrating via the circular case.)

These four pitch categories amount to the auditory atoms of a mover's trajectory. Given the sequence of Doppler pitches of a mover, it is easy to decompose it into the fundamental atoms of movement the mover is engaged in. Let's walk through these four kinds of pitch profiles, and the four respective kinds of movement they indicate, keeping our eye on Figure 28.

(A) The *bottom right* square in Figure 28 shows a situation where the pitch is low and rising. Low pitch means my neighborhood ice cream truck is directed away from me and the kids, but the fact that the pitch is rising means the truck is turning and directing itself more toward us. Intuitively, then, a low and rising pitch is the signature of an away-moving mover noticing you and deciding to begin to turn around and come see you. To my snack-happy children, it means hope—the ice cream truck might be coming back!

(B) The *upper right* square concerns cases where the pitch is higher than baseline and is rising. The high pitch means the truck is directed at least somewhat toward us, and the fact that the pitch is rising means the truck is further directing itself toward us. Intuitively, the truck has seen my kids and is homing in on them. My kids are ecstatic now, screaming, "It's coming! It sees us!"

(C) The *top left* square is where the pitch is still high, but now falling. That the pitch is high means the truck is headed in our direction; but the pitch is falling, meaning it is directing itself less and less toward us. "Hurry! It's here!" my kids cry. This is the signature of a mover *arriving*, because when movers arrive at your destination, they either veer away so as not to hit you, or come to a stop; in each case, it causes a lowering pitch, moving toward baseline.

(D) The *bottom left, and final,* square of the matrix is where the pitch is low and falling. This means the truck is now directed away, and is directing itself even *farther* away. Now my kids' faces are purple and drenched with tears, and I am preparing a plate of carrots.

Figure 28 amounts to a second kind of ecological pitch–movement dictionary (in addition to Figure 26). Now, if melodic pitch contours have been culturally selected to mimic Doppler shifts, then the dictionary categorizes four fundamentally different meanings for *melody*. For example, when a melody begins at the bottom of the pitch range of a piece and rises, it is interpreted by your auditory system as an away-moving mover veering back toward the listener (bottom right of Figure 28). And if the melody is high in pitch and falling, it means the fictional mover is arriving (upper left of Figure 28). At least, that's what these melodic contours mean if melody has been selected over time to mimic Doppler shifts of movers. With some grounding in the ecological meaning of pitch, we are ready to begin asking whether signatures of the Doppler effect are actually found in the contours of melody. We begin by asking how many fingers one needs to play a melody.

## ONLY ONE FINGER NEEDED

Piano recitals for six-year-olds tend to be one-finger events, each child wielding his or her favorite finger to poke out the melody of some nursery rhyme. If one didn't know much about human music and had only been to a kiddie recital, one might suspect that this is because kids are given especially simple melodies that they can eke out with only one finger. But it is not just kindergarten-recital melodies that can be played one note at a time, but nearly all melodies. It appears to be part of the very nature of melody that it is a strictly sequential stream of pitches. That's why, even though most instruments (including voice, for the most part) are capable of only one note at a time, they are perfectly able to play nearly any melody. And that's also why virtually every classical theme in Barlow and Morgenstern's *Dictionary of Musical Themes* has just one pitch at a time.

Counterexamples to this strong sequential tendency of melody are those pieces of music having two overlapping melodies, or one melody overlapping itself, as in a round or fugue. But such cases serve as counterexamples that prove the rule: they are *not* cases of a single melody relying on multiple simultaneous notes, but, rather, cases of *two* simultaneously played single melodies, like the sounds of *two* people moving in your vicinity.

Could it be that melodies are one note at a time simply because it is physically difficult to implement multiple pitches simultaneously? Not at all! Music revels in having multiple notes at a time. You'd be hard put to find music that does not liberally pour pitches on top of one another—but not for the melody.

Why is melody like this? If chords can be richly complex, having many simultaneous pitches, why can melodic contour have only one pitch at a time? There is a straightforward answer if melodic contour is about the Doppler pitch modulations due to a mover's direction relative to the listener. A mover can only possibly be moving in a *single* direction at any given time, and therefore can have only a single Doppler shift relative to baseline. Melodic contour, I submit, is one pitch at a time because movers can only go in one direction at a time. In contrast, the short-time-scale pitch modulations of the chords are, I suggested earlier in the chapter, due to the pitch constituents found in the gangly bangs of human gait, which *can* occur at the same time. Melodic contour, I am suggesting, is the Doppler shifting of this envelope of gangly pitches.

## HUMAN CURVES

Melodic contours are, in the sights of this movement theory of music, about the sequence of movement directions of a fictional mover. When the melody's pitch changes, the music is narrating to your auditory system that the depicted mover is changing his or her direction of movement. If this really *is* what melody means, then melody and people should have similar *turning* behavior.

How quickly *do* people turn when moving? Get on up and let's see. Walk around and make a turn or two. Notice that when you turn 90 degrees, you don't usually take ten steps to do so, and you also don't typically turn on a dime. In order to get a better idea of how quickly people tend to change direction, I set out to find videos of people moving and changing direction. After some thought, undergraduate RPI student Eric Jordan and I eventually settled on videos of soccer players. Soccer was perfect because players commonly alter their direction of movement as the ball's location on the field rapidly changes. Soccer players also exhibit the full range of human speeds, allowing us to check whether turning rate depends on speed. Eric measured 126 instances of approximately right-angle turns, and in each case, recorded the number of steps the player took to make the turn. Figure 29 shows the distribution for the number of steps taken. As can be seen in the figure, these soccer players typically took two steps to turn 90 degrees, and this was the case whether they were walking, jogging, or running. Casual observation of movers outside of soccer games—such as in coffee shops—suggests that this is not a result peculiar to soccer.

If music sounds like human movers, then it should be the case that the depicted mover in music turns at rates typical for humans. Specifically, then, we expect the musical mover to take a right-angle turn in about two steps on average. What does this mean musically? A step in music is a beat, and so the expectation is that music will turn 90 degrees in about two beats. But what does it mean to "turn 90 degrees" in music?

Recall that the maximum pitch in a song means the mover is headed directly toward the listener, and the lowest pitch means the mover is headed directly away. *That* is a 180-degree

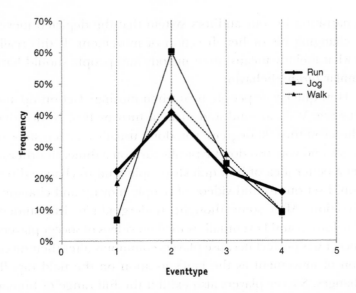

**FIGURE 29**. Distribution of the number of footsteps soccer players take to turn 90 degrees, for walkers, joggers, and runners. The average number of footsteps for a right angle turn for walkers is 2.16 (SE 0.17, n=22); for joggers, 2.21 (SE 0.11, n=45); and for runners is 2.23 (SE 0.13, n=59).

difference in mover direction. Therefore, when a melody moves over the entirety of the *tessitura* (the melody's pitch range), it means that the depicted mover is changing direction by 180 degrees (either from toward you to away from you, or vice versa). And if the melody spans just the top or bottom half of the tessitura, it means the mover has turned 90 degrees. Because human movers take about two steps to turn 90 degrees—as we just saw—we expect that melodies tend to take about two beats to cross the upper or lower half of the tessitura.

To test this, we measured from the *Dictionary of Musical Themes* melodic "runs" (i.e., strictly rising or strictly falling sequences of notes) having at least three notes within the upper or lower half of the tessitura (and filling at least 80 percent of the width of that half tessitura). These are among the clearest potential candidates for 90-degree turns in music. Figure 20 shows how many beats music typically takes to do its 90-degree turns. The peak is at two beats, consistent with the two footsteps of people making 90 degree turns while moving. Music turns as quickly as people do!

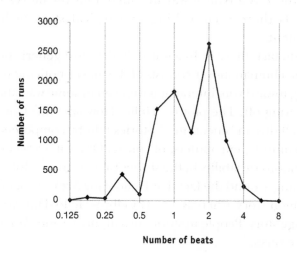

**FIGURE 30.** Distribution of the number of beats for a theme "run" to cover the top or bottom half of the tessitura. The most common number of beats to cross half the tessitura is two, consistent with how quickly people turn 90 degrees. (Runs had three or more notes moving in the same direction, and all within the top or bottom half of the tessitura, and filling at least 80% of its half tessitura.)

This close fit between human and melodic turning rates is, as we will see, helpful in trying to understand regularities in the overall structure of melody, which we delve into next.

## MUSICAL ENCOUNTERS

In light of the theory that music is a story about a fictional mover in our midst, is it possible to address what a typical melody is like? How do melodies usually begin? What is the typical melodic contour shape? How many beats are in a typical melody? Is there even such a thing as a "typical" melody?

To answer whether there is a typical melody, we must ask if there is a typical way in which a mover moves in our midst. Is there such a thing as a typical story of human movement?

Yes, in fact, there is. In a nutshell, the most generic possible story of a human mover consists of the mover noticing you and veering toward you, interacting with you in some way, and then scampering off. The plot is: "Hello. How are you? Goodbye." Let's call this an "encounter." Stories told by composers by no means must be encounters, of course. One can expect to find tremendous variability in the stories told by music. But there is no getting around the fact that Hello–HowAreYou–Goodbye is a common story, whereas, say, Goodbye–Hello–HowAreYou is a strange story. People in stories usually arrive and *then* depart, not vice versa.

If "encounters" are indeed the most typical story of human movement in our midst, then let's be more specific about the kinds of movement involved. Figure 31 shows the four qualitatively distinct kinds of movement we discussed earlier in the chapter. Can we say what an encounter is in terms of these movement categories? Yes. In the "Hello" part of the encounter, the mover suddenly notices you and begins veering toward you. This is case A or B in Figure 31, depending on whether the mover was receding or nearing, respectively, when he first noticed you. If A is the start of the story, then B occurs next. By the end of B, the mover has gotten near enough that he must begin veering away, lest he bump into you. This is the segment of the mover's path that brings him closest to you—where he says, "How are you?"—and is case C in Figure 31. Finally, in the "Goodbye" part of the encounter, the mover veers away, which is case D.

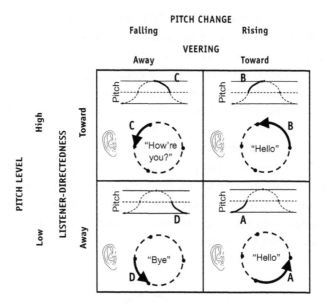

**FIGURE 31.** Summary of the movement meaning of pitch for low and high, rising and falling pitch, as shown earlier in Figure 17. The sequence A-B-C-D is a plausible "most generic" movement, which I call an "encounter," or Hello–HowAreYou–Goodbye movement.

The generic encounter, then, has one of two sequences, B-C-D or A-B-C-D, with the movement meanings of A through D defined in Figure 21. Furthermore, I suggest that the "full" A-B-C-D movement is the more common of the two, because stories of human movement often consist of multiple, repeated encounters, and in these cases A will be the first segment of any encounter after the first. That is to say, if a mover goes away, then turns to come back for another encounter, the sequence must begin at A, not B. We conclude, then, that the generic Hello–HowAreYou–Goodbye movement is thus A-B-C-D. A-B is the "Hello". C is the "How are you?" And D is the "Goodbye."

We now have some idea of what a generic movement is, and so the next question is, "What does it sound like?" With Figure 31 in hand, we can immediately say what the Doppler pitches are for this movement. Figure 32 illustrates the overhead view and the pitch contour of the generic A-B-C-D encounter. This generic pitch contour in Figure 32 has two distinctive features that we expect, no matter the specific shape of the mover's encounter path. First and foremost, the generic pitch contour goes up, and then down. Second, it dwells longer (has a flatter slope) at the minimum and maximum of its pitch range (something I will discuss in the Encore section titled "Home Pitch"). In essence, this pitch contour shares two qualitative features that are crucial to *hills*: hills are not just lumps of earth on the ground—not domes, not mounds, not pyramids, not wedges—but have gentle slopes toward their bottoms and a gentle flattening at the top.

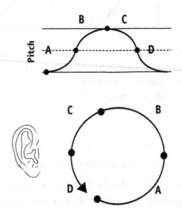

**FIGURE 32.** The most generic sequence of movement, and its pitch over time. One can see it is a "hill." Furthermore, this path tends to be traversed in around eight steps (because people tend to take about two steps to turn 90 degrees).

We are nearly ready to ask whether melodies tend to look like the "hill" we see in the pitch contour of the generic encounter, but first we must gather one further piece of information about encounters. We need to know how long

an encounter lasts. When movers do a Hello–HowAreYou–
Goodbye to us, do they do it over 100 footsteps, or four? Our
results from the previous section ("Human Curves") can help
us answer. We saw then that human movers tend to take about
two steps to turn 90 degrees. Each of the four segments of the
generic encounter—A, B, C, and D—is roughly a 90-degree
turn, and thus the entire encounter can be carried out in
about eight steps. Although Hello–HowAreYou–Goodbye
behaviors can take fewer or more than eight footsteps, a
plausible baseline expectation is that generic encounters
will occupy around eight steps—not two steps, and not 80.
The generic story of human movement in our midst—the
"encounter"—is not just the sequence of movements A-B-C-
D, but also, these movements being enacted in eight or so
steps. Accordingly, the generic Doppler pitch contour is not
just a hill, but a hill implemented in about eight footsteps.
The sound of a generic human mover in our midst is an eight-
step pitch hill.

These eight-step-hill encounters are, I claim, a fundamental
intermediate-level structure found in the pattern of Doppler
pitches from human movers in our midst. Real movers will,
of course, often diverge from this, but such instances should
be viewed as deviations from this generic baseline. I call the
generic encounter an *intermediate-level* structure because it is
hierarchically above the individual "atoms" of movement (A,
B, C, and D), and because *full* stories of movers in our midst
may involve hundreds or thousands of footsteps—full stories
of human movement are built by combining many short bouts
of movement. Because the encounter is the generic short bout
of movement, the generic long story of human movement is
many eight-step hills—many encounters.

And *now* we are in a position to figure out whether these
generic stories of human movement are found in music. In particular, we want to know if melodies are built out of encounter-
like structures. And because generic encounters sound like

eight-step pitch hills, we wish to see if melodies have any tendency to be built out of eight-beat pitch hills. Of course, we expect that any such tendency should be weak: the eight-step pitch hill is the expectation for the generic encounter, but you can be sure that composers like to tell non-generic stories as well. Nevertheless, we hope to see the signs of the generic melody by looking across a great many melodies.

Eric Jordan and I set out to measure the average pitch contour for themes, following the lead of Professor David Huron, who first carried out measurements of this kind and found arches in average pitch contours. Themes were put into groups with other themes having the same number of notes; each theme's pitches were normalized (so that the bottom and top of the tessitura were 0 and 1, respectively); and the average normalized pitch was computed across the group. Classical themes tend to have fewer than 25 notes, and in order to sample longer melodies, allowing us to better discern signs of eight-beat hills, we also measured from a set of 10,000 Finnish folk songs, which have themes with longer lengths of 25 to 40 notes. Figure 33 shows the average pitch contours for each group. There are, for example, 83 themes having exactly eight notes (the average of their contours is shown at the upper left in Figure 33). If melodies are built from eight-beat hills, as predicted, then we expect to see such hills in these averaged melodies. A casual glance across the average pitch contours for melodies having the same number of notes indicates a multiple-hill pattern in longer melodies. For themes with eight to 13 notes, only one hill is apparent in the average contour, but by 14 through about 19 notes, there is an apparent bimodality to the plots. Among the plots with 30 or more notes, a multiple-hill contour is strongly apparent. And the hills are very roughly eight notes in length. (These eight-note hills are due to themes with a predominance of quarter notes. Themes with a preponderance of eighth notes lead to 16 note hills.)

## AVERAGE MELODIC PITCH CONTOUR

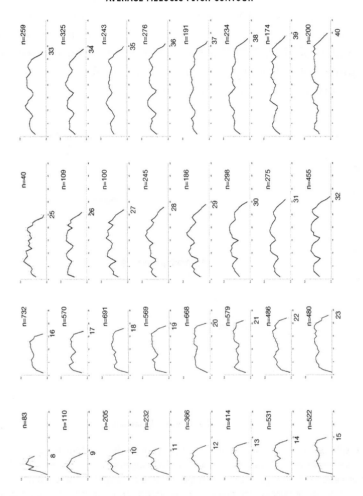

**FIGURE 33**. Average melodic contour for melodies with same number of notes. Thirty two different plots are shown, for melodies having eight notes through 40 notes, shown by the number label along each x-axis. The "n" values show the number of melodies over which the average is calculated. If melodies are built with eight-beat hills, then because a sizable fraction of melodies consist mostly of beat-long notes, there should be a strong eight-note-hill tendency in these plots. There will also be sizable fraction of melodies consisting mostly of half-beat-long notes, and these will tend to have a sixteen-note hill. The smallest hills we expect to see, then, are eight-note ones.

To get a more quantitative estimate of the typical number of notes per hill in these average-melodic-contour data, Figure 34 plots the approximate number of hills as a function of the number of notes in the average melody. One can see that the number of hills is approximately ⅛ the number of notes, and so there are about eight notes per hill.

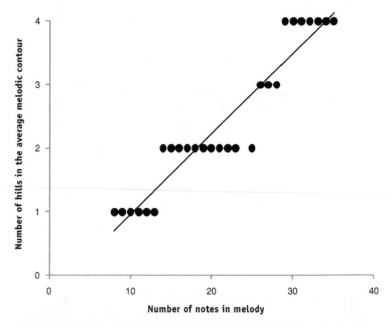

**FIGURE 34.** Number of hills in the average melodic contour as a function of the number of notes in the melody. The approximate number of arches from each plot in Figure 19 having eight notes through 36 notes (after which the number of hills is less clear) was recorded. The number of hills rises as 0.1256 times the number of notes in the melody. The number of notes per hill is thus ⅟.1256, which is 7.90, or approximately 8, consistent with our expectation from generic human encounters.

The data we have just discussed provide evidence of the "eightness" of melody's hills. But recall that in order to show that they are *hills*, and not some other protuberance shape, we must show that the note durations at the start, peak, and end of the protuberance tend to have longer duration, just as hills are flatter on their ends and on top (see Figure 32 again). Focusing now just on classical themes with eight, nine, and ten notes, Eric Jordan and I determined the average duration (in beats) of each note. To generate a hill-shaped pitch contour, we would expect a "W"-shaped plot of how note durations vary over the course of a melodic theme, with longer durations at the start, middle, and end. Indeed, that's what we found, as shown in Figure 35. Melodies tend to have longer-duration notes at the start and end (when the fictional mover is headed away from the listener), and also in the middle (when the mover is headed directly toward the listener).

Melodies appear to have a tendency to be built out of 8-beat hills, which is just what is expected if melody's stories are about a fictional mover's multiple encounters with the listener. Like arrival, interaction, and departure in eight steps, melodies tend to rise and fall over eight beats, and in the nonlinear fashion consistent with a real encounter. Melodies, in other words, seem to have the signature structure of stories built from Hello–HowAreYou–Goodbyes.

In this and the previous sections we have seen that melody behaves in some respects like the Doppler pitch modulations of a person moving. But there's much more to the similarity between Doppler and melody, and I discuss additional similarities in detail in the Encore. Here I will only hint at them, but I encourage you to read the hints, because they are exciting, and they are crucial to the case I am making.

**Encore 3: "Fancy Footwork"** This section will discuss how when people turn and their Doppler pitch changes, their gait can often become more complex. And I will provide evidence that music behaves in the same way. I referred to Encore 3 earlier in this chapter when we wrapped up the discussion of rhythm, because "Fancy Footwork" concerns how rhythm and pitch interact.

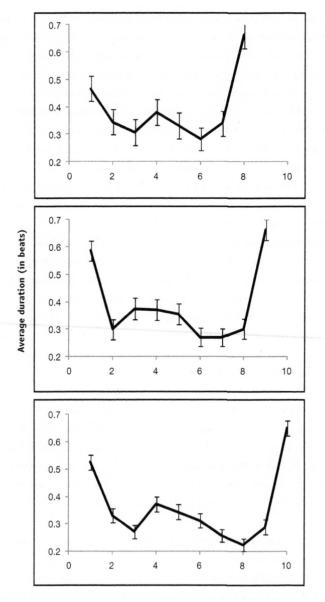

**FIGURE 35.** Average duration in beats of each note for classical themes with number of notes near that of the generic encounter (i.e., for 8, 9 and 10 notes). One can see the expected "W" shape, showing that the eight-note-hills in Figure 23 are truly *hills*.

**Encore 5: "Home Pitch"** This Encore section will discuss three similarities between melody and Doppler pitch modulations, each concerning how pitch distributes itself. Like Doppler pitch modulations, melodic contours have a fixed home range (called the tessitura); tend to distribute themselves uniformly over their home range; and tend to dwell longer at the edges of the pitch range.

**Encore 6: "Fast Tempo, Wide Pitch"** Just as faster movers have wider Doppler pitch ranges, faster-tempo music has a wider pitch range.

**Encore 7: "Newton's First Law of Music"** If melodic contours are Doppler pitch modulations, then we expect a variety of asymmetries between pitch rises and falls. Pitch changes are not generally expected to have "momentum" (i.e., a tendency to continue going in the same direction) with the exception of small downward changes in pitch. Also, melody is expected to have a tendency to drift gradually down, but to have larger upswings in pitch.

## WHERE'S THE MOVING PITCH?

Throughout this chapter thus far, I have been suggesting that the Doppler effect is the ecological foundation of melody. One potential stumbling block for this hypothesis that I have until now avoided mentioning is that we're not generally aware of these Doppler shifts in pitch. We may sometimes consciously notice them coming from skateboards, bikes, cars, and trains, but do we notice them coming from moving *people*? If our auditory system is listening to the Doppler shifts of people, then wouldn't we have noticed? Not necessarily. As we discussed in Chapter 1, in the section titled "Under the Radar," your conscious self not only doesn't acknowledge these lower-level stimuli; it typically does not even have access to them. Just as your conscious self sees objects, not visual contour junctions, your conscious self hears the behaviors and actions of movers, not the auditory substrate out of which the whole is built. Your conscious self tends to latch onto the invariants out there—a pink room with a variety of colored lights will have a variety of spectra, but you'll typically see the walls as uniformly pink. Similarly, when a mover in your midst is making sounds, your conscious self will focus on invariants such as *what* the mover is doing. Your auditory system will hear *all* of the lower-level structure, including the (detectable) pitch modulations due to the Doppler Effect, but all that will tend to stay below the radar because *you* don't need to know the details.

It would not, then, be surprising if our auditory system detects (and uses) Doppler shifts from human movers and yet we don't consciously notice it. But one might wonder whether our auditory system can possibly detect Doppler shifts for movers going human speeds in the first place. If I am walking at one meter per second, then the difference in pitch between the highest and lowest pitch in the Doppler range is small, roughly about a tenth of a semitone on the piano. The fastest sprints possible by humans are about 10 meters per second, and even at these speeds the Doppler pitch range is only about a single semitone on the piano (e.g., from C to C#). Thus, whereas the Doppler pitch modulations for trains, planes, and automobiles are sizable because those vehicles move quickly, the pitch modulations for humans are small. Tiny as these Doppler pitch shifts may be, however, our auditory system is exquisitely sensitive to pitch changes, capable of detecting differences as small as about 0.3 percent of the frequency, or about 5 percent of a semitone—sensitive enough to distinguish even the pitch shifts of walkers.

So our ears are sensitive enough. But if one thinks more carefully about how our feet make sounds, there appears to be a big problem with the suggestion that the sounds of human movers make Doppler shifts at all. The biggest bang your body makes in motion is when your foot takes a step. But notice that when you are walking and your foot hits the ground, it is moving neither forward nor backward relative to the ground. Your feet are *stationary* when they touch the ground. But if your sound-making foot is not moving forward when it makes its sound, then its pitch will not get Doppler shifted upward. That is, even though *you* are moving steadily forward, your feet are *standing still* at the moments when they are making their footstep sounds, and so your footstep sounds undergo no Doppler shift. (See Figure 36a.) (This is not a problem that applies to between-the-steps gangly bangings, which *do* Doppler shift.)

Footsteps are, however, more subtle than a simple vertical thud on the ground, and there are multiple avenues in which Doppler shifts may occur. First, even if your footsteps stomp the ground without any forward speed, your *body* is still moving forward. When the footstep sound waves rise into the air, your big old body bumps into them and reflects them. The waves that your front runs into get reflected forward and are consequently Doppler upshifted, and the waves that reflect off your back get Doppler downshifted, as illustrated in Figure 36b (i). Second, footsteps aren't the simple vertical ground bangers that Figure 36a would have us believe. For one thing, the dynamics of a footstep are complicated by the fact that the *ground* is often complicated. Throughout most of our evolutionary history the ground was not smooth as it is today, but covered in grass, brush, and pebbles. When your foot goes in for a landing, it has some forward velocity, and will undergo a sequence of micro-collisions with the ground material. This sequence of collisions is a forward-moving sequence, and will sound higher in pitch (or compressed) to a listener toward whom the mover is directed. (See Figure 36b [ii].) Not only is the ground more complicated than the "stomping" in Figure 36a indicates, but the foot dynamics even on *smooth* ground are more subtle than I have let on. Our heel hits first, and the contact points move forward along the foot toward the toes (Figure 36b [iii]). Whether on smooth ground or natural terrain, this sequence of microhits underlying a single step is moving in the direction of the mover, and thus will Doppler shift.

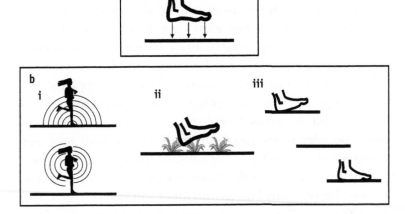

FIGURE 36. (a) When a foot hits the ground, it is not moving forward or backward, and therefore has no Doppler shift. But as we'll discuss, there's more to the story. (b) (i) Top: The footstep leads to sound waves going in all directions, all at the same frequency (indicated by the spacing between the wave fronts). Bottom: These waves hit the body and reflect off it. Because the mover is moving forward, the sound waves reflected forward will be Doppler shifted to a higher pitch; the waves hitting the mover's rear will reflect at a lower pitch. (ii) When feet land they don't simply move vertically downward for a thud. The surface of the ground very often has complex material on it which the landing foot stikes as it is still moving forward. These complex sounds will have Doppler shifts. (iii) If our feet were like a pirate's peg leg, then the single thud it makes when hitting the ground would have no Doppler shift. But our feet aren't peg legs. Instead, our foot lands on its heel, and the point of contact tends to move forward along the bottom of the foot.

Footsteps can, then, Doppler shift, and these shifts are detectable. There is now a third difficulty that can be raised: if Doppler shifts for human movers are fairly meager, then why doesn't musical melody have meager tessitura width (i.e., meager pitch range for the melody)? The actual tessitura in melody tends to be wider than that achievable by a human mover, corresponding to speeds faster than humans can achieve. Why, if melodic pitch contours are about Doppler pitches, would music exaggerate the speed of the depicted observer? Perhaps for the same reason that exaggeration is commonplace in other art forms. Facial expressions in cartoons, for example, tend to be hyperexaggerations of human facial expressions. Presumably such exaggerations serve as superstimuli, hyperactivating our brain's mechanisms for detecting the characteristic (e.g., smile or speed), and something about this hyperactivation feels good to us (perhaps a bit like being on a roller coaster).

One final thought concerning the mismatch between the Doppler pitch range and the tessitura widths found in music: could I have been underestimating the size of the Doppler shifts humans are capable of? Although we may only move in the one- to ten-meters-per- second range, and our limbs may swing forward at a little more than twice our body's speed, *parts of us* may be moving at faster speeds. Recall that your feet hit the ground from heel to toe. The sequence of microhits travels forward along the bottoms of your feet, and the entirety of sound the sequence makes will be Doppler shifted. An interesting characteristic of this kind of sound is that it can act like a sound-making object that is moving much faster than the object that actually generates it. As an example, when you close scissors, the actual objects—the two blades—are simply moving toward each other. But the *point of contact* between the blades moves *outward* along the blades. The sound of closing scissors is a sound whose principal source location is moving, even though no object is actually moving in that manner. This kind of faux-moving sound maker can go very fast. If two flattish surfaces hit each other with one end just ever so slightly ahead of the other, then the speed of the faux mover can be huge. For example, if you drop a yardstick so as to make it land flat, and one end hits the ground one millisecond before the other end, then the faux mover will have traveled between the yardstick and the ground from one end to the other at about one kilometer per second, or about two thousand miles per hour! The faux mover beneath our stepping feet may, in principle, be moving much faster than we are, and any scissor-like sound it makes will thus acquire a Doppler pitch range much wider than that due to our body's natural speed.

Human movers *do* make sounds that Doppler shift, and these shifts are detectable by our auditory system. And their exaggeration in music is sensible in light of the common role of exaggeration in artistic forms. Melodic contour, we have seen thus far, has many of the signature properties expected of Doppler shifts, lending credence to the idea that the role of melodic pitch contours is to tell the story of the sequence of directions in which a mover is headed. That's a fundamental part of the kinematic information music imparts about the fictional mover. But that's only half the story of "kinemusic." It doesn't tell us how far away the mover is, something more explicitly spatial. That is the role of loudness, the topic of the rest of this chapter.

## LOUD AND IN 3-D

Do you know why I love going to live shows like plays or musicals? Sure, the dialog can be hilarious or touching, the songs a hoot, the action and suspense thrilling. But I go for another reason: the 3-D stereo experience. Long before movies were shot and viewed in 3-D, people were putting on real live performances, which provide a 3-D experience for all the two-eyeds watching. And theater performances don't simply approximate the 3-D experience—they are the genuine article.

"But," you might respond, "one goes to the theater for the dance, the dialog, the humans— for the art. No one goes to live performances for the '3-D feel!' What kind of lowbrow rube are you? And, at any rate, most audiences sit too far away to get much of a stereo 3-D effect."

"Ah," I respond, "but that's why I sit right up front, or go to very small theater houses. I just *love* that 3-D popping-out feeling, I tell ya!"

At this point you'd walk away, muttering something about the gene pool. And you'd be right. That *would* be a dopey thing for me to say. We see people doing their thing in 3-D all the time. I just saw the waitress here at the coffee shop walk by. Wow, she was in 3-D! Now I'm looking at my coffee, and my mug's handle appears directed toward me. Whoa, it's 3-D!

No. We don't go to the live theater for the 3-D experience. We get plenty of 3-D thrown at us every waking moment. But this leaves us with a mystery. Why *do* people like 3-D movies? If people are all 3-D'ed out in their regular lives, why do we jump at the chance to wear funny glasses at the movie house? Part of the attraction surely is that movies can show you places you have never been, whether real or imaginary, and so with 3-D you can more fully experience what it is like to have a *Tyrannosaurus rex* make a snout-reaching grab for you.

But there is more to it. Even when the movie is showing everyday things, there is considerable extra excitement when it is in 3-D. Watching a live performance in a tiny theater is still not the same as watching a 3-D movie version of that same performance. But what is the difference?

Have you ever been to one of those shows where actors come out into the audience? Specific audience members are sometimes targeted, or maybe even pulled up on stage. In such circumstances, if you're *not* the person the actors target, you might find yourself thinking, "Oh, that person is having a blast!" If you're the shy type, however, you might be thinking, "Thank God they didn't target me because I'd have been terrified!" If you *are* the target, then, whether you liked it or not, your experience of the evening's performance will be very different from that of everyone else in the audience. The show reached out into *your* space and grabbed *you*. While everyone else merely watched the show, you were part of it.

The key to understanding the "3-D movie" experience can be found in this targeting. 3-D movies differ from their real-life versions because *everyone* in the audience is a target, all at the same time. This is simply because the 3-D technology (projecting left- and right-eye images onto the screen, with glasses designed to let each eye see only the image intended for it) gives everyone in the audience the *same* 3-D effect. If the dragon's flames appear to me to nearly singe my hair but spare everyone else's, your experience at the other side of the theater is that the dragon's flames nearly singe *your* hair and spare everyone else's, including mine. If I experience a golf ball shooting over the audience to my left, then the audience to my left also experiences the golf ball going over *their* left. 3-D movies put on a show that is inextricably tied to each listener, and invades each listener's space equally. Everyone's experience is identical in the sense that they're all treated to the same visual and auditory vantage point. But everyone's experience is unique because each experiences *himself* as the target—each believes he has a specially targeted vantage point.

The difference, then, between a live show seen up close and a 3-D movie of the same show is that the former pulls just one or several audience members into the thick of the story, whereas 3-D movies have this effect on *everyone*. So the fun of 3-D movies is not that they are 3-D at all. We can have the same fun when we happen to be the target in a real live show. The fun is in being *targeted*. When the show doesn't merely leap off the screen, but leaps at *you*, it fundamentally alters the emotional experience. It no longer feels like a story about others, but becomes a story that invades your space, perhaps threateningly, perhaps provocatively, perhaps joyously. You are immersed in the story, not an audience member at all.

What does all this have to do with music and the auditory

sense? Imagine yourself again at a live show. You hear the performers' rhythmic banging ganglies as they carry out behaviors
on stage. And as they move on stage and vary their direction,
the sounds they make will change pitch due to the Doppler
effect. Sitting there in the audience, watching from a vantage
point *outside* of the story, you get the rhythm and pitch modulations of human movers. You get the attitude (rhythm) and
action (pitch). But you are not immersed in the story. You can
more easily remain detached.

Now imagine that the performers suddenly begin to target
you. Several just jumped off the stage, headed directly toward
you. A minute later, there you are, grinning and red-faced, with
tousled hair and the bright red lipstick mark of a mistress's
kiss on your forehead . . . and, for good measure, a pirate is in
your face calling you "salty." During all this targeting you hear
the gait sounds and pitch modulations of the performers, but
you also heard these sounds when you were still in detached,
untargeted audience-member mode. The big auditory consequence of being targeted by the actors is not in the rhythm
or pitch, but in the *loudness.* When the performers were on
stage, most of the time they were more or less equidistant, and
fairly far away—and so there was little loudness modulation as
they carried on. But when the performers broke through the
"screen," they ramped up the volume. It is these high-loudness
parts of music—the fortissimos, or *ff*s—that are often highly
evocative and thrilling, as when the dinosaur reaches out of
the 3-D theater's screen to get you.

And that's the final topic of this chapter: loudness, and
its musical meaning. I will try to convince you that loudness
modulations are used in music in the 3-D, invade-the-listen-
er's-space fashion I just described. In particular, this means
that the loudness modulations in music tend to mimic loudness modulations due to changes in the *proximity* of a mover.
Before getting into the evidence for this, let's discuss why I
don't think loudness mimics something *else.*

## NEARNESS VERSUS STOMPINESS

I will be suggesting that loudness in music is primarily driven by spatial proximity. Rather than musical pitch being a spatial indicator, as is commonly suggested (see the earlier section, "Why Pitch Seems Spatial"), it is *loudness* in music that has the spatial meaning. As was the case with pitch, here, too, there are several stumbling blocks preventing us from seeing the spatial meaning of loudness. The first is the bias for pitch: if one mistakenly believes that pitch codes for space, then loudness must code for something else. A second stumbling block to interpreting loudness as spatial concerns musical notation, which codes loudness primarily via letters (*pp, p, mf, f, ff,* and so on), rather than as a spatial code (which is, confusingly, how it codes pitch, as we've seen). Musical instruments throw a third smoke-screen over the spatial meaning of loudness, because most instruments modulate loudness not by spatial modulations of one's body, but by hitting, bowing, plucking, or blowing *harder.*

Therefore, several factors are conspiring to obfuscate the spatial meaning of loudness. But, in addition, the third source of confusion I just mentioned suggests an alternative interpretation: that loudness concerns the energy level of the sound maker. A musician must use more energy to play more loudly, and this can't help but suggest that louder music might be "trying" to sound like a more energetic mover. The energy with which a behavior is carried out is an obvious real-world source of loudness modulations. These energy modulations are, in addition, highly informative about the behavior and expressions of the mover. A stomper walking nearby means something different than a tiptoer walking nearby. So energy or "stompiness" is a potential candidate for what loudness might mean in music.

Loudness in the real world can, then, come both from the *energy* of a mover and from the *spatial proximity* of the mover. And each seems to be the right sort of thing to potentially explain why the loudest parts of music are often so thrilling and evocative: stompiness, because the mover is energized (maybe angry); proximity, because the mover is very close by. Which of these ecological meanings is more likely to drive musical loudness, supposing that music mimics movement? Although I suspect music uses high loudness for both purposes—sometimes to describe a stompy mover, and sometimes to describe a nearby mover—I'm putting my theoretical money on spatial proximity.

One reason to go with the spatial-proximity interpretation of loudness, at the expense of the stompiness interpretation, is pragmatic: the theory is easier! Spatial proximity is simply distance from the listener, and so changes in loudness are due to changes in distance. *That's* something I can wrap my theoretical head around. But I don't know how to make predictions about how walkers vary in their stompiness. Stompers vary their stompiness when *they* want to, not in the way physics wants to. That is, if musical loudness is stompiness, then what exactly does this predict? It depends on the psychological dynamics of stompiness, and I don't *know* that. So, as with any good theorist, spatial proximity becomes my friend, and I ignore stompiness.

But there is a second reason, this one substantive, for latching onto spatial proximity as the meaning of musical loudness. Between proximity and stompiness, proximity can better explain the large range of loudness that is possible in music. Loudness varies as the *inverse square* of proximity, and so it rises dramatically as a mover nears the listener. Spatial proximity can therefore bring *huge* swings in loudness, far greater than the loudness changes that can be obtained by stomping softly and then loudly at a constant distance from a listener. That's why I suspect proximity is the larger driver of loudness modulations in music. And as we will see, the totality of loudness phenomena in music are consistent with proximity, and less plausible for stompiness (including the phenomenon discussed in Encore 5, that note density rises with greater loudness).

Thus, to the question "Is it nearness or stompiness that drives musical loudness modulations?" the answer, for both pragmatic and substantive reasons, is nearness, or proximity. Nearness can modulate loudness much more than stompiness can, and nearness is theoretically tractable in a way that stompiness is not. Let's see if proximity can make sense of the behavior of loudness in music.

## SLOW LOUDNESS, FAST PITCH

Have you ever wondered why our musical notation system is as it is? In particular, why does our Western music notation system indicate pitch by shifting the notes up and down on the staff, while it indicates loudness symbolically by letters (e.g., *pp*, *f*) along the bottom? Figure 37 shows a typical piece of music. Even if you don't read music—and thus don't know exactly which pitch each note is on—you can instantly interpret how the pitch varies in the melody. In this piece of music, pitch rises, wiggles, falls, falls, falls yet again, only to rise and tumble down. You can *see* what pitch does because the notation system creates what is roughly a plot of pitch versus time. Loudness, on the other hand, must be read off the letters along the bottom, and their meaning unpacked from your mental dictionary: *p* for "quiet", *f* for "loud", and so on. Why does pitch get a nice mapping onto spatial position, whereas loudness only gets a lookup table, or glossary?

**FIGURE 37.** The usual notation, where vertical position indicates pitch, and intensities are shown along the bottom. The music is a simplification of the 7th through 12th measures from Johann Christoph Friedrich Bach's *Menuet and Alternativo*. It utilizes standard music notation. Standard notation is sensible because pitches vary much more quickly than loudness, so it tends to not be a problem to have to read the loudness levels along the bottom.

Music notation didn't *have* to be like this. It could do the reverse: give loudness the spatial metaphor, and relegate pitch to being read along the bottom in symbols. Figure 38 shows the same excerpt we just saw in Figure 37, but now in this alternative musical notation system. Going from the lowest horizontal line upward, the lines now mean *pianissimo* (*pp*), *piano* (*p*), *mezzo forte* (*mf*), *forte* (*f*), and *fortissimo* (*ff*). The pitches for each note of the song are now shown along the bottom. Once one sees this alternative notation system in Figure 38, it becomes obvious why it is a terrible idea. When vertical height represents loudness, vertical height tends to just stay constant for long periods of time. The first eight notes are all at one loudness level (*piano*), and the remaining twelve are all at a second loudness level (*forte*). Visually there are just two plateaus, severely underutilizing your visual talents for seeing spatial wiggles. In standard notation, where pitch is spatially represented, on the other hand, the notes vary vertically much more on the page. Not only does our hypothetical alternative notation underutilize the capabilities of the visuospatial code, it overutilizes the letter codes. We end up with "word salad" along the bottom. In this case, there are 15 instances where the pitch had to be written down, nearly as many as there are notes in the excerpt. In standard notation, where *loudness* is coded via letters, there were just *two* letters along the bottom (see Figure 37 again).

**FIGURE 38.** A notation system in which vertical position indicates intensity, and pitches are shown along the bottom. This is the same excerpt from J.C.F Bach as in Figure 37 but here the notation schemes for pitch and loudness have been swapped. This is *not* a good way to notate music because most of the note-to-note modulations are in pitch, not in loudness, which means an overabundance of pitch labels along the bottom, and little use of the vertical dimension within the horizontal bars. (To read the pitches along the bottom, "E5", for example, is the E pitch at the fifth octave. When the octave is not labeled, it is presumed to be that of the last mentioned octave.)

---

The reason this alternative music notation system is so bad is, in one sense, obvious. In music, pitch typically varies quickly, often from note to note. Loudness, on the other hand, is much less variable over time. As exemplified by the excerpt in Figure 37, pitch can change at very short time scales, such as a 16th or 32nd note, but intensities typically persist for *many measures* before they change. To illustrate this, consider the fictional piece of music shown in Figure 39 (using standard music notation). In this example, pitch hardly ever changes, and loudness is changing very quickly. Music is never like this, which is why the standard notation system is a good one. The standard music notation system is so useful for music because it gives the quickly varying musical quality—pitch—the spatial metaphor, and relegates to the glossary the musical quality that stays much more constant: loudness.

**FIGURE 39.** An alternative kind of music that never happens (shown in regular notation). If music were often like this, then the alternative notation system in Figure 38 *would* be sensible. (This fictional music was created by taking J.C.F. Bach's piece in Figure 32, keeping the notes as in that reverse notation system, but then pretending they represent pitches, as in normal notation. ("*f4*" means "*ffff*".)

To get a more quantitative measure of how quickly pitches and loudnesses change over time, Sean Barnett measured the distribution of time spent on a pitch before changing (Figure 9, earlier), and Eric Jordan measured the distribution of time spent at a given level of loudness before changing (Figure 40 below). One can see that pitches typically switch after about half a beat to a beat, whereas intensities change after about 10 beats.

While it is obvious why the standard notation system is smarter than my hypothetical "reverse" one for music, it is *not* obvious why music is like this in the first place. Why does music

*have* "fast pitches" and "slow loudnesses"? If music sounds like
movement, and loudness modulations are selected to sound
like those due to spatial proximity, then the answer is straightforward. In order to significantly change loudness, the mover
must move some distance through space. Melodic pitch, on
the other hand, is about directedness toward the listener. In
contrast to movement through space, which takes a relatively
long time, a mover can "turn on a dime." In a single step, a
mover can turn about 45 degrees with ease (and typically
does), which would translate to a fourth of the tessitura width
(on average). Melodic pitch changes more quickly than loudness in music simply because human movers can change direction more quickly than they can change their proximity to the
listener. That's why music never sounds like the fictional piece
of music in Figure 33, and that's why our Western musical
notation system is an efficient one. The comparative time

scales of loudness and melodic pitch are what we should expect if music sounds like human movers, with loudness modulated by the spatial proximity and melodic pitch by the direction of the depicted mover.

In addition to the time scale for loudness modulations being consistent with that for changes in proximity, additional evidence for proximity as the meaning of loudness is provided in two Encore sections:

**Encore 4: "Distant Beat"** I will discuss how the nearer movers are, the more of their gait sounds are audible, and how this is also found in music: louder portions of music tend to have more notes per beat. (This was also mentioned earlier in this chapter as we finished up our discussion of rhythm, because it concerns the interaction between rhythm and loudness.)

**Encore 6: "Fast Tempo, Wide Pitch"** I discuss how, as expected from theory, music with a faster tempo has a wider pitch range for its melody. This Encore section also shows, however, that— as predicted—the range of *loudnesses* in a piece is *not* correlated with tempo.

**Encore 7: "Newton's First Law of Motion"** This Encore section takes up a variety of predictions related to the inertia of moving objects, on the one hand, and the asymmetry between pitch rises and pitch falls, on the other. We will predict, and data will confirm, that this asymmetry changes as a function of loudness: when music indicates (by high loudness level) that the mover is close, the probability rises of long pitch runs downward.

**Encore 8: "Medium Encounters"** This Encore section concerns regularities in how movers distribute themselves in distance from a listener, and makes predictions about how frequently music makes use of various loudness levels.

## SUMMARY

In this chapter and the previous chapter, we have covered a great deal of musical ground (and we will cover still more in the Encore chapter). In Chapter Three, we presented general arguments for the music-is-movement theory, clearing three of the four hurdles for a theory of music: why we have a brain for music, why music should be emotionally moving, and why music should elicit movements in us. In this chapter, we have addressed the fourth hurdle, explaining the structural features of music. As I hope readers can see, there is a wealth of suspicious similarities between music and the sounds of people moving—42 suspicious similarities—which are summarized in the table below.

| SECTION | HUMAN MOVERS | MUSIC |
|---|---|---|
| **1. Drum Core ( p. )** | Footsteps are regularly repeating. | The beat is regularly repeating. |
| **2. Drum Core (p. )** | Footsteps are the most fundamental auditory feature of human movement. | The beat is the most fundamental quality of music. |
| **3. Drum Core (p. )** | Footsteps tend to be around 1 to 2 per second. | Beats tend to be around 1 to 2 per second. |
| **4. Drum Core (p. )** | Footsteps usually are not as regular as a metronome. | Beats are often looser than that of a metronome. |
| **5. Drum Core (p. )** | People's footstep rates lower prior to stopping (*ritardando*). | The number of beats per second lowers prior to musical endings (*ritardando*). |
| **6. Gangly Notes ( p. )** | Footsteps are usually higher-energy collisions than between-the-step bangs. | On-beat notes usually have greater emphasis than off-beat notes. |
| **7. Gangly Notes (p. )** | In addition to footsteps, people's gangly limbs make sounds in between the footsteps. | In addition to notes on the beat, music has notes in between the beats. |
| **8. Gangly Notes (p. )** | The between-the-steps gangly bangs are time-locked to the steps. | The between-the-beats notes are time-locked to the beat. |

| SECTION | HUMAN MOVERS | MUSIC |
|---------|--------------|-------|
| 9. Gangly Notes (p. ) | The pattern of steps and between-the-steps gangly bangs is crucial to identifying the mover's behavior. | The pattern of on-beat and off-beat notes (the rhythm) is crucial to the identity of a song. |
| 10. Gangly Notes (p. ) | Human-mover gait sounds (steps and between-the-steps banging ganglies) have rings, and often pitches. | Musical notes often have pitches. |
| 11. The Length of Your Gangly (p. ) | People typically make about 0 or 1 between-the-step bang. | Music typically has about one off-beat note per beat. |
| 12. Backbone (p. ) | Footsteps can be highly variable in intensity, and we perceptually sense a step even when inaudible. | Beats are felt even when no note occurs on a beat. |
| 13. Backbone (p. ) | It is not merely the temporal pattern of gait sounds that identifies a mover's behavior. It matters which sounds are on the beat. | The feel of a musical rhythm does not depend solely on the temporal pattern, but on where the listener interprets the beats to be. |
| 14. The Long and Short of Hit ( p. ) | People are likely to make a between-the-steps gangly bang near the middle of a step cycle. | Off-beat notes most commonly occur near the middle of a beat cycle. |
| 15. The Long and Short of Hit (p. ) | People are more likely to make a between-the-steps gangly bang just before a step than just after. ("Long-shorts" are more common.) | Off-beat notes more commonly occur in the second half of a beat cycle (just before the beat) than in the first half (just after the beat). |
| 16. Measure of What? (p. ) | Patterns of footstep emphases are informative as to the mover's behavior. | Time signature matters to the identity of music. |
| 17. Gangly Chords (p. ) | Gait sounds have temporal patterns *and* pitch patterns (due to the pitches of the constituent ganglies). | Music typically has rhythm chords. |
| 18. Gangly Chords (p. ) | A mover's temporal pattern of hits is matched to the pitch pattern (because the pitches are due to the constituent gangly bangs). | Chords (e.g., as played with the left hand on the piano) have the same time signature as the rhythm. |

| SECTION | HUMAN MOVERS | MUSIC |
|---------|--------------|-------|
| 19. Gangly Chords (p. ) | Footsteps tend to have lower pitch than other gangly bangs. | For chords, the pitch played on the beat tends to be lower than that played off the beat. |
| 20. Gangly Chords (p. ) | The pitches among gangly bang sounds can occur simultaneously (unlike Doppler shifts, see below). | Chords are often struck simultaneously. |
| 21. Fancy Footwork (p. ) | When people turn, they tend to have more complex gangly bangings. | When melodic contour rises or falls, the rhythm tends to be more complex. |
| 22. Distant Beat (p. ) | People that are nearer have more audible gangly bangs per step. | Louder music has more notes per beat. |
| 23. Choreographed for You (p. ) | Doppler shift pitch contours and loudness contours matter for the appropriate visual-auditory fit of a human mover. | Melodic contours and loudness contours (not just rhythm) are relevant for choreographers in creating of visual movements to match music. |
| 24. Why Pitch Seems Spatial (p. ) | Doppler pitches change continuously over time. | Melodic contour tends to change fairly continuously. |
| 25. Only One Finger Needed (p. ) | A mover is only moving in one direction at any moment, and thus has only one Doppler shift for a listener. | Melodies are inherently one pitch at a time. |
| 26. Home Pitch (p. ) | For a mover at constant speed, Doppler shifts are confined to a fixed range, the highest (lowest) corresponding to heading directly toward (away from) the listener. | Melodies tend to be confined to a fixed range of pitches called the tessitura. |
| 27. Home Pitch (p. ) | People tend to move in all directions relative to a listener, and to fairly uniformly sample from Doppler pitches within the Doppler range. | Melodies tend to sample fairly uniformly across their tessitura. |
| 28. Home Pitch (p. ) | Pitches at the top and bottom of the Doppler range tend to have longer duration (due to trigonometry). | Melodies tend to have longer-duration notes when the pitch is at the top or bottom of the tessitura. |

| SECTION | HUMAN MOVERS | MUSIC |
|---|---|---|
| **29. Fast Tempo, Wide Pitch** (p. ) | Faster movers have a wider Doppler pitch range. | Faster tempo music tends to have a wider tessitura. |
| **30. Fast Tempo, Wide Pitch** (p. ) | Faster movers do *not* have a wider range of proximity-based loudnesses. | Faster tempo music does *not* tend to have a wider range of loudness. |
| **31. Human Curves** (p. ) | People take about two steps to make a right angle turn. | Music takes about two beats to traverse the top or bottom half of the tessitura (which corresponds to a right-angle turn). |
| **32. Musical Encounters** (p. ) | The most generic kind of human encounter is the Hello–HowAreYou–Goodbye, involving a circling movement beginning when a mover headed away begins to turn toward the listener. The Doppler pitch contour is that of a hill, with flatter slopes at the bottom and top. | Melodies in music have a tendency to be built out of pitch hills. |
| **33. Musical Encounters** (p. ) | The generic encounter tends to be around 8 steps (2 steps per right-angle turn). | The constituent pitch hills in melodies tend to be roughly eight beats long. |
| **34. Newton's First Law of Music** (p. ) | Changing Doppler pitches have little or no tendency to continue changing, consistent with Newton's First Law of Motion (inertia). | When melodic contour varies, there is little or no tendency to continue changing. |
| **35. Newton's First Law of Music** (p. ) | More subtly, Doppler shifts possess "momentum" only when falling by a small amount. | More subtly, melodic contours possess "momentum" only when falling by a small amount. |
| **36. Newton's First Law of Music** (p. ) | Small Doppler pitch changes are more likely downward, and large pitch changes more likely upward. | Small melodic contour changes are more likely downward, and large changes more likely upward. |
| **37. Newton's First Law of Music** (p. ) | Extended segments of falling Doppler pitch are more common than extended segments of rising Doppler pitch (due to passing movers). | Downward melodic runs are more common than upward melodic runs. |

| SECTION | HUMAN MOVERS | MUSIC |
|---|---|---|
| **38. Newton's First Law of Music (p. )** | More proximal, and thus louder, movers are more likely to undergo large downward Doppler pitch runs. | Louder portions of music are more likely to feature large pitch runs downward (compared to upward). |
| **39. Slow Loudness, Fast Pitch (p. )** | People can turn quickly, and can thus change Doppler pitch quickly (i.e., half a tessitura in about two steps). But people cannot typically change loudness quickly, because that requires actually moving many steps across space. | Melodic contour changes quickly, but loudness changes at much slower time scales. |
| **40. Medium Encounters (p. )** | Encounters with a person have an average distance, spending more total time at near-average distances than at disproportionately near or far distances (in contrast to the fairly uniform distribution of Doppler pitches). | Most pieces have a typical loudness level (e.g., *mezzo forte*), spending most of their time at that loudness level, and progressively less time at loudness levels deviating more from this average (in contrast to the fairly uniform distribution of melodic pitches). |
| **41. Medium Encounters (p. )** | In any given encounter, a person is more commonly more distant than average than more proximal (because there's more "far" real estate than "near"). | The distribution of times spent at each loudness level is not only peaked, but asymmetrically disfavors the louder loudness levels. |
| **42. Medium Encounters (p. )** | Nearer-than-average portions of a mover's encounter tend to be shorter in duration than farther-than-average portions. | Louder-than-average segments of music tend to be more transient than softer-than-average segments. |

# So What Are We?

You can't catch a cat with a carrot. Unless you fashion the carrot into the spear tip of a harpoon and have good aim, you'll have better luck luring the cat with tuna fish. Even though cats didn't evolve eating 500-pound tuna, the meaty odor and taste taps into what cat noses and tongues like. If you keep up a daily dose of fish—and a wee bit of water—you're likely to have caught yourself a cat. Add a litter box next to your toilet and this cat will quickly become no more likely than you to stain the carpet. With these two simple steps, and only $39 in supplies, you'll have quickly transformed a wild animal with hundreds of millions of years of evolutionary design into a toilet-trained, self-cleaning vermin remover.

Unlike dogs, who have changed their brains and bodies for hundreds of thousands of years to suit man—that is, they've become domesticated—cats are notoriously *not* domesticated.

Cats may be our occasional friends and usefully serve as clean critter catchers, but *they* think they are wild cats, and they do what wild cats tend to do. They fit into our lives not because they have evolved for us, but because we've shaped our houses—and it doesn't take much shaping—so that cats behaving like cats naturally leads to a function fulfilled for us.

Cats are *harnessed*, not domesticated—their innate talents redirected, with little or no training, in ways evolution didn't intend. And the fundamental trick behind our harnessing cats is so simple we hardly appreciate it; in fact, I've already mentioned it. Tuna, not carrots, and kitty litter, not a bidet. Although tuna is not what cat ancestors ate, tuna is sufficiently meat-ish in odor and taste that it fits right into a cat's finicky diet disposition. And although kitty litter is a strange and unnatural concoction, it mimics the loose soil in which cats prefer to bury their feces. Tuna and kitty litter are simulacra of nature, and thus successfully harness cats, turning them into good pets.

Just as we find cats with lifestyles they are not meant to have, we humans are apes living an un-ape life. Far beyond being potty trained, we build the very toilets we sit on. In light of tuna fish and kitty litter, in this book we have examined whether simulacra of nature could be the key to our humanity. Rather than supposing that our wild bits have evolved and changed to help us become modern humans, and rather than supposing the opposite—that our feral brain is a general-purpose learning machine that is extensively wired during our lifetime to make us genteel—in this book we have examined a third possibility: that our brains, to this day, are just as they were before anyone spoke or folded napkins, and that culture evolved to harness our ape powers, cleverly turning them into a new kind of power. Apes became literate and musical not because language and music got themselves innately encoded into the brain, but because the brain got its signature stamped upon language and music. We're cats, not dogs.

In particular, this book has been about culture's general strategy for harnessing us. The trick is to structure modern human tasks as tasks at which our ape selves already excel. And one sure-fire way to do this is to make the task thoroughly "like nature." This book set out to put flesh on the bones of this idea, and to convey preliminary evidence that this *is* the strategy culture used to make us fit into modernity: making modernity fit us.

(a) Speech sounds like solid-object physical events, (b) music sounds like people moving, and (c) *Homo sapiens* became modern human by virtue of cultural evolution designing language and music to mimic nature—by virtue of nature-harnessing. That's the book in a nutshell. We have been down and dirty discussing (a) and (b) over the last three chapters, and these, in conjunction with arguments in *The Vision Revolution* that writing looks like opaque objects strewn in 3-D space, are the principal arguments for (c), that nature-harnessing is the mechanism for how we got from ape to man. Speech and music are the most central and transformative—revolutionary—powers we possess, and that's why it is reasonable to say that nature-harnessing is the mechanism that created humans.

And if nature-harnessing is what made us, and if other animals don't get made in this way, then it is worth reflecting upon what we humans *are*. How can we think about ourselves? To help illustrate what we are, it helps to back up really far away. So let's consider how aliens try to make sense of us. . . .

When aliens come to Earth to investigate life here, they don't simply beam up a specimen and start probing. (And they're also, by the way, not disproportionately interested in the anus.) Only a novice prober would do a simple beam-and-probe, and would surely get a quick rap on the proboscis from the instructor. The problem with abducting an animal of interest, all by itself, is that you can't understand an animal without an appreciation of the environment the animal inhabits.

What an experienced alien prober does is gather as much information about the animal's habitat as possible. In fact, the aliens beam up entire habitats so that they can study the animal in its home at their leisure. Alien Probe School graduates are consummate ecologists, understanding that organisms evolved to do stuff with their complex mechanisms, but that if you drop an organism into an environment for which it did not evolve, it will often do other stuff, and usually quite unsophisticated stuff.

By following their alien principles of good probing, they'll have abducted what they need in order to someday, and with great effort, have a thorough knowledge of the organism, from its genome to its "phenome." The phenome is the set of things the animal can do, implemented ultimately by the genome and the way it acts within the evolutionary habitat. For example, your cell phone's genome is its electronic circuitry (or perhaps the engineer's drawings for the circuitry), whereas its phenome is the list of things it can do, often enumerated in the user's manual—exactly the manual that is missing for the Earth organisms the alien probers want to unravel.

But something unexpected happened when they applied these wise principles to humans, abducting an entire primitive tribe and the mountain they lived on. They already had abducted earlier hominids who had no language or music, and were interested to see what was new about these speaking and singing humans. To their surprise, the aliens could discern no difference between the nonspeaking, nonmusical hominids and the speaking and singing humans. Their biology was indistinguishable, they concluded. They were the same animal. Could the difference be due to a difference in habitat? No, they concluded, the earlier and newly abducted mountains appear to have no relevant differences. Same animal, same habitat, and yet the modern humans are a giant leap beyond, or at least distinct from, the more ancient *Homo sapiens*.

They scratched their antennae. Why, the aliens wondered,

did the modern humans behave so fundamentally differently? Why did they have language and music? Why did the modern humans seem like something fundamentally different from the great apes, whereas the nonlinguistic, nonmusical humans seemed to fit more within the apes, albeit as a very bright great ape? How could two identical creatures in identical habitats end up so different in sophistication that it seemed natural to deem them different species?

The modern humans clearly must have *learned* language and music. But that only created another dilemma for the probers. How can you teach an animal a lesson so powerful that it practically becomes another species? Speech and music comprehension, the aliens knew, are astoundingly complex, with just the kind of complexity natural selection creates. These modern humans, the aliens noted, were competent at language and music in the highly adapted way animals evolve to be good at things. But from their alien experiences as ecologists, they knew that if an animal is not designed to accommodate that level and type of complex processing, then you can't just force-feed it the learning. You can't teach a deer to catch and eat mice. No training course will get your dog to climb trees like a monkey. And you can't train a human to comprehend fax machine sounds. You simply cannot teach old hominids new tricks worthy of natural selection. The human brain is not such a rich general-purpose learning apparatus that it can master tasks as richly complex as language and music. Yet there they were, modern humans with brains highly honed for speech and music. The alien probers were stumped.

They reasoned that humans don't have language or music innately installed in their heads. And neither one comes from their natural habitat. And they also can't simply learn something that complicated. There must be selection of some kind underlying the human capability to do language and music, but what kind of selection could it be, if it is neither natural selection nor learning?

One of the alien probers wondered whether there *might* be design, or selection, underlying the difference between modern humans and their nonlinguistic and nonmusical ancestors—not natural selection, but *cultural* selection. This is a selection process that selects not on biology, but on human artifacts that are used by biology. The human artifacts are animal-like in the sense that they themselves have evolved over time, under selection pressure. These artifact-creatures (in the realm of "memes"), like naturally selected biological creatures, can be highly complex and intricately adapted, with all the hallmarks of an engineering masterpiece.

"Aha!" the alien prober exclaimed. The modern humans are not merely learning language and music, they're being raised in an environment with symbionts. Language and music are technological masterpieces that evolved to live with nonlinguistic hominids and transform them into something beyond their biology. What makes these modern humans no longer the nonlinguistic *Homo sapiens* apes they biologically are is not on the inside, and not in the ancestral natural environment. Language and music are evolved, organism-like artifacts that are symbiotic with these human apes. And like any symbiont, these artifact symbionts have evolved to possess shapes that fit the partner biology—our brains.

What are we, then, in the eyes of alien probers? We are our biology, from the genes on up. But we are more than that, as indicated by the fact that the probers don't abduct just a human, but, rather, abduct entire human habitats. We are our biology within its appropriate habitat. But that's true of all animals on Earth. The special thing the aliens had to grapple with when they started probing humans was that biology and habitat are not enough. They needed to abduct the cultural-artifact symbionts that were coevolving with us. That's not something any other animal can lay claim to. The pieces of what we are can be found in our wet biology, and in the habitat, but also in the artifactual symbionts we have been coevolving with. Our language, music, and other highly culturally evolved technologies are, like our genes and our habitat, deeply part of the modern human recipe. The human code is not just the genome, and not just the genome plus the habitat. The human code is now partly found in the structures of language and other cultural artifacts.

Through this allegory of alien probers, we can better see what we are. We owe our modern human identity to cultural symbionts which have evolved to get into our brains and harness us into something new. Cultural "animals" evolving to be symbiotic with humans: *that* is something the aliens could wrap their proboscises around, for they knew of lots of symbiotic interactions around the galaxy.

Although the aliens concluded that these cultural symbionts must have culturally evolved to fit the human brain, they hadn't figured out how the symbionts got *into* human heads. "How did the cultural symbionts get in?" they wondered. And the answer they discovered: "Ah! They got in by mimicking nature."

# Encore

Although Chapter 4 presented a variety of evidence that the structure of music has the signature of human movers, there is additional evidence that couldn't reasonably be fit into that chapter, and so it appears here in the Encore.

## 1 THE LONG AND SHORT OF HIT

The mysterious approaching monster from the section titled "Backbone" in Chapter 4 was mysterious because you mistakenly perceived a hit sound rapidly following the footstep; that is, you perceived the between-the-steps interval to be split into a short interval (from step to rapidly following hit) and a long interval (from that quick post-step hit to the next footstep). The true gait of the approaching lilting lady had its between-the-steps interval broken, instead, into a long interval followed by a short interval. My attribution of mystery to the "short-long" gait, not the "long-short," was not arbitrary. "Short-long" *is* a strange human gait pattern, whereas "long-short" is commonplace.

Your legs are a pair of twenty-five-pound pendulums that swing forward as you move, and are the principal sources of your between-the-steps hit sounds. A close look at how your legs move when walking (see Figure 1) will reveal why between-the-step hits are more likely to occur just *before* a foot-step than just after. Get up and take a few steps. Now try it in slow motion. Let your leading foot hit the ground in front of you for its step. Stop there for a moment. *This* is the start of a step-to-step interval, the end of which will occur when your now-trailing foot makes *its* step out in front of you. Before continuing your stride, ask yourself what your trailing foot is doing. It isn't doing anything. It is on the ground. That is, at the start of a step-to-step interval, *both* your feet are planted on the ground. Very slowly continue your walk, and pay attention to your trailing foot. As you move forward, notice that your trailing foot stays planted on the ground for a while before it eventually lifts up. In fact, your trailing foot is still touching the ground for about the first 30 percent of a step-to-step interval. And when it finally does leave the ground, it initially has a very low speed, because it is only just *beginning* to accelerate. Therefore, for about the first third of a step, your trailing foot is either not moving or moving so slowly that any hit it does take part in will not be audible. Between-the-footsteps hit sounds are thus relatively rare immediately after a step. After this slow-moving trailing-foot period, your foot accelerates to more than *twice* your body speed (because it must catch up and pass your body). It is during this central portion of a step cycle that your swinging leg has the energy to really bang into something. In the final stage of the step cycle, your forward-swinging leg is decelerating, but it still possesses considerable speed, and thus is capable of an audible hit.

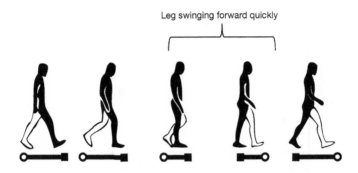

Leg swinging forward quickly

FIGURE 41. Human gait. Notice that once the black foot touches the ground (on the left in this figure), it is not until the next manikin that the trailing **(white)** foot lifts. And notice how even by the middle figure, the trailing foot has just begun to move. During the right half of the depicted time period, the white leg is moving quickly, ready for an energetic between-the-steps hit on something.

We see, then, that there is a fundamental temporal asymmetry to the human step cycle. Between-the-steps hits by our forward-swinging leg are most probable at the middle of the step cycle, but there is a bias toward times nearer to the later stages of the cycle. In Figure 41, this asymmetry can be seen by observing how the distance between the feet changes from one little human figure to the next. From the first to the second figure there is no change in the distance between the feet. But for the final pair, the distance between the feet changes considerably. For human gait, then, we expect between-the-steps gangly hits as shown in Figure 42a: more common in mid-step than the early or late stages, and more common in the late than the early stage.

Does music show the same timing of when between-the-beat notes occur? In particular, are between-the-beat notes most likely to occur at about the temporal center of the interval, with notes occurring relatively rarely at the starts and ends of

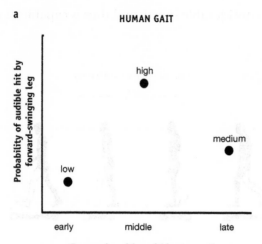

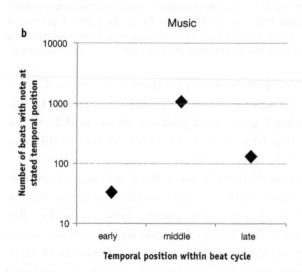

FIGURE 42. (a) Because of the nature of human gait, our forward-swinging leg is most likely to create an audible between-the-steps bang near the middle of the gait cycle, but with a bias toward the late portions of the gait cycle, as illustrated qualitatively in the plot. (b) The relative commonness of between-the-beat notes occurring in the first half ("early"), middle, or second half ("late") portions of a beat cycle. One can see the qualitative similarity between the two plots.

the beat cycle? And, additionally, do we find the asymmetry that off-beat notes are more likely to occur late than early (i.e., are long-shorts more common than short-longs)? This is, indeed, a common tendency in music. One can see this in the classical themes as well, where I measured intervals from the first 550 themes in Barlow and Morgenstern's dictionary, using only themes in 4/4 time. There were 1078 cases where the beat interval had a single note directly in the center, far more than the number of beat intervals where only the first or second half had a note in it. And the gait-like asymmetry was also found: there were 33 cases of "short-longs" (beat intervals having an off-beat note in the first half of the interval but not the second half, such as a sixteenth note followed by a dotted eighth note), and 131 cases of "long-shorts" (beat intervals having a note in the second half of the interval but not the first half, like a dotted eighth note followed by a sixteenth note). That is, beat intervals were four times more likely to be long-short than short-long, but both were rare compared to the cases where the beat interval was evenly divided. Figure 42b shows these data.

Long-shorts are more common in music because they perceptually feel more natural for movement—because they *are* more natural for movement. And, more generally, the time between beats in music seems to get filled in a manner similar to the way ganglies fill the time between steps. In the Chapter 4 section titled "The Length of Your Gangly," we saw that beat intervals are filled with a human-gait-like *number* of notes, and now we see that those between-the-beat notes are positioned inside the beat in a human-gait-like fashion.

Thus far in our discussion of rhythm and beat, we have concentrated on the temporal pattern of notes. But notes also vary in their emphasis. As we mentioned earlier, on-beat notes typically have greater emphasis than between-the-beat notes, consistent with human movers typically having footsteps more energetic than their other gangly bangs. But even notes on the beat vary in their emphasis, and we take this up next.

## 2 MEASURE OF WHAT?

Thus far we have discussed beats as footsteps, and between-
the-beat notes as between-the-footsteps banging ganglies. But
there are other rhythmic features of music that occur at the
scale of *multiple* beats. In particular, music rarely treats each
and every beat as equal. Some beats are special. In ¾ time,
for example, every third beat gets a little emphasis, and in ⁴⁄₄
time every fourth beat gets an emphasis. This is the source
of the *measure* in music, where the first beat in each measure
gets the greatest emphasis. (And there are additional patterns: in ⁴⁄₄ time, for instance, the third beat gets a little extra
oomph too, roughly half that of the first.) If you keep the
notes of a piece of music the same, but modify which beats
are emphasized, the song can often sound nearly unrecog-
nizable. For example, here is "Twinkle, Twinkle Little Star,"
but with some unusual syllables emphasized to help you sing
it in ¾ time rather than the appropriate ⁴⁄₄ time. "*TWI*-nkle,
twi-**NKLE**, lit-tle *STAR*, <silent beat>, how *I* won-der *WHAT*
you are." As you can see, it is very challenging to even get
yourself to sing it in the wrong time signature. And when you
eventually manage to do it, it is a quite different song from
the original.

Why should a difference in the pattern of emphasis on
beats make such a huge difference in the way music sounds
to us? With the movement theory of music in hand, the question becomes: does a difference in the pattern of emphasis
of a mover's footsteps make a big difference in the meaning
of the underlying behavior? For example, is a mover with a ¾
time gait signature probably doing a different behavior than a
mover with a ⁴⁄₄ time gait signature?

The answer is, "Of course." A different pattern in footstep
emphasis means the mover is shifting his body weight in a different pattern. The ¾ time mover has an emphasis on every

third step, and thus alternates which foot gets the greater emphasis. The $\frac{4}{4}$ time mover, on the other hand, has emphasis on every other step, with extra emphasis on every fourth step. These are the gait sounds of distinct behaviors. Real movements by people may not stay within a single time signature for prolonged period, as music often does, but, instead, change more dynamically as the mover runs, spins, and goes up for a layup. Time-signature differences in movement imply differences in behavior, and so we expect that our auditory system is sensitive to these time signatures . . . and that music may have come to harness this sensitivity, explaining why time signature matters in music.

And notice that when we hear music with a time signature, we want to move consistently not only with the beat and the temporal pattern of notes, but *also* with the time signature. People *could* waltz to music with a $\frac{4}{4}$ time signature, but it just does not feel right. People not only want to step to the beat, (something we discussed early in Chapter 4); they want to step extra hard on the emphasized beat.

This and the previous Encore section concerned rhythm. The upcoming two also concern rhythm, and how it interacts with melody and with loudness, respectively.

## 3 FANCY FOOTWORK

When the kids and I are doing donuts in the parking lot at the dollar store—that is, driving the minivan in such tight circles that the wheels begin to screech and squeal—we are making minivan gangly banging sounds. Such behavior leads to especially complex rubber-meets-road hits and slides, sounds we revel in as we're doing it. But the patrons at the dollar store hear an additional feature. The patrons hear Doppler shifts, something that the kids and I do not hear because we are stationary relative to the minivan. For the dollar store patrons, the pitch of the envelope of minivan gangly bangings rises and falls as we approach and recede from them in our donuts. In fact, it is *because* my minivan is veering so sharply that its ganglies begin banging in a more complex fashion. Compared to minivans *not* doing donuts, minivans doing donuts change pitch faster *and* have more complex "gaits." Greater pitch changes therefore tend to be accompanied by more complex gait patterns.

This pitch-rhythm connection is also found among human movers. When we turn, we are likely to have a more complex gait and gangly pattern than when we are simply moving straight ahead. For example, when you turn left, you must lean left, lest you fall over on your right side; and your legs can no longer simply swing straight past each other, but must propel the body leftward via a push or pivot. And many turns involve more complex footwork, such as sidestepping, trotting, twists, and other maneuvers we acrobatic apes regularly carry out. For example, when a basketball player crosses the court, his or her path is roughly straight, and the resultant gait sounds are a simple beat. Once a player crosses the court, however, his or her movements tend to be curvy, not straight, as players on offense try to free themselves up for a pass, or players on defense loom in for a steal or shadow the offense to prevent a pass, in each case setting off a richer pattern of gangly sounds.

Does music behave in this way? When melodic pitches change—a signal that the depicted mover is turning, as we discussed in Chapter 4—does the rhythm tend to get more complex? As a test for this, I sampled 713 two-beat intervals having at least two notes each from the *Dictionary of Musical Themes*, and for each recorded whether the pitch was varying or unvarying, and whether the rhythm was simple (one note on each beat, or "just the footsteps") or complex (more than "just the footsteps"). (Data were sampled from $2/4$ and $4/4$ time signature pieces, and from every tenth theme up to "D400" in the Dictionary.) When pitch changed over the two-beat intervals, the probability was 0.66 that the beat was complex, whereas when pitch did not change the probability was only 0.35 that the beat was complex. Consistent with the prediction from real-world turners, then, these data suggest that when music changes pitch—the Doppler signature of a mover changing direction—its rhythm tends to become more complex. That is, as with people, when music "turns," the ganglies start flying.

We see, then, that melody interacts with rhythm in the way Doppler interacts with gait. Now let's ask whether loudness also interacts with rhythm, as expected from the ecology of human movers. We take that up in the next Encore section.

## 4 DISTANT BEAT

As I write this I am on the (inner) window ledge of my office at RPI, overlooking downtown Troy and the Hudson River. I'm on the fifth floor (of the city side of the building), with a steep, sloping hill at the bottom, so everything I hear is either fairly far away, or very far away. Because of my extreme distance from nearly everything, I end up hearing only a small sample of the sounds occurring in the city. Mainly, I hear the very energetic events. If a sound were not very energetic, then it would be inaudible by the time the sound waves reached me. I can see a tractor dumping rocks, but I hear only the boom of a particularly large one, missing out on the sounds of the many smaller rock hits I can see but cannot hear. Generally, when something makes complex sounds, whether it is a car, a washing machine, or a tornado, some of the noises composing the whole are more energetic than others. If it is far away from you, then you will only hear the most energetic parts of the sound. But if you are close, you'll be able to hear the full panoply of sounds.

As with most complex sound makers, human movers make sounds of varying energy and frequency. The most energetic sounds tend to be our footsteps. Accordingly, the first thing we hear when someone is approaching from afar tends to be their footsteps. The other gait-related sounds from clanging limbs are difficult to hear when far away, but they get progressively more audible as the mover nears us. That is, as a mover gets closer to us and the loudness of his gait sounds thereby rises, the number of audible gait sounds per footstep tends to increase.

If music has culturally evolved to sound like human movement, then we accordingly expect that the louder parts of songs should have more notes per beat (i.e., more fictional gangly bangs per step). Do they? Do *fortissimo* passages have greater note density than *pianissimo*? Caitlin Morris, as an undergraduate at RPI, set out to test this among scores in *The Classical Period: An Anthology of Piano Music, Vol. II*, by Denes Agay (New York: Music Sales America, 1992), and found that this is indeed the case. Figure 43 shows how the density of notes (the number of notes per beat) varies with loudness over 60 classical pieces. One can see that note density increases with loudness, as predicted. Music doesn't have to be like this. Music *could* pack more notes in per beat in soft parts, and have only on-the-beat notes for the loud parts. Music has this louder-is-denser characteristic because, I submit, *that's* a fundamental *ecological* regularity our auditory systems have evolved to expect for human (and any) movers.

This result is, by the way, counter to what one might expect if loudness were due not to spatial proximity but to the energy level (or "stompiness," as we discussed in the Chapter 4 section titled "Nearness versus Stompiness") of the mover. Louder stomps typically require longer gaps between each stomp. "Tap, tap, tap, tap, tap" versus "BANG! . . . . . . . . BANG!"

Now that we have expanded on rhythm, we will move on to further evidence that melodic contour acts as Doppler pitch.

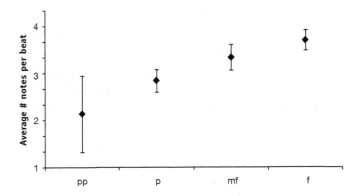

**FIGURE 43.** Data from 60 pieces in Denes Agay's *An Anthology of Piano Music, Vol. II: The Classical Period* showing that louder portions of music tend to be packed with more notes. Each of 234 contiguous segments of constant loudness were sampled, counting the total number of notes and beats; averages are over these 234 segments. Data collected and analyzed by Caitlin Morris. (Standard errors shown.)

# 5 HOME PITCH

We have discussed how melody consists of one pitch at a time (in the Chapter 4 section titled "Only One Finger at a Time"), which is exactly what we expect if melody has its origins in Doppler shifts. We also mentioned earlier in that chapter (in the section titled "Why Pitch Seems Spatial") that melodic pitch tends to change in a fairly continuous fashion. At any one time, then, melody is at one pitch, and changes pitch roughly as if it is "moving" through pitches. Thinking of melody as if it is an unknown creature we wish to better understand, it is natural to ask about melody's home and where it roams within the space of pitches. In this Encore section we'll discuss three facets of melody's home range and foraging behavior.

Let's begin with one of the most salient features about the Doppler effect, which is that for a mover at constant speed there is a maximum and minimum pitch the mover can attain, these occurring when the mover is headed directly toward and directly away from the listener, respectively. Doppler pitches for any mover are therefore bound to a fixed home range. If Doppler pitches are confined in this sense to a fixed home range, then the music-is-movement theory predicts that melodies, too, should tend to confine themselves to a fixed home range. Melodies should tend to behave as if there is an upper or lower boundary to pitch. *Does* melody move around as if bound within an invisible fence, as predicted, or does melody move more freely? Although melody is highly variable, it has long been noticed that any given melody tends to confine itself to a fairly fixed window of pitches called its tessitura. The notion of tessitura allows that the melody may occasionally punch through a barrier, but the barriers are still worthy of recognition because of their tendency to hold the pitch inside. The tessitura is, I submit, music's implicit recognition that a single constant-speed mover has a fixed range of Doppler pitches it can express.

Melody, then, has a fixed home range, consistent with Doppler shifts. Let's now look into how melody spends its time within its home range. If melody really is acting like Doppler shifts, then melody should distribute itself within its home range in a similar manner to Doppler pitches. How *do* Doppler pitches distribute themselves within their home range? In particular, for a mover in your vicinity carrying out behaviors, in which directions does the mover tend to go? Many of the movers around you are just doing their own thing, carrying out actions that do not involve the fact that you are there. These movers will tend to go in any direction relative to you. But even movers who are interacting in some way with you will tend not to strongly favor some directions over others. For example, performers on a stage will, over the course of the show, move in all directions relative to any audience member. In fact, although their actions onstage may be highly intricate, they will often be very roughly summarized as moving in circles out in front of the listener, an illustrative case we had used earlier in Figure 25 of Chapter 4. Such circle-like behavior tends to sample broadly from all directions. Individual short bouts of behavior, then, can be anywhere in the Doppler pitch range. *Across* bouts of behavior, then, Doppler pitches tend to occur with fairly uniform probability over the Doppler range. For melody, then, we expect that individual melodic themes will be highly variable in their pitch distribution, but we also expect that, on average, these themes will sample pitches within their tessitura fairly uniformly.

That is, in fact, what we found among the classical themes. Despite wide variability from theme to theme, across the 10,000 classical themes the average distribution of notes across the tessitura is fairly flat, as shown in Figure 44. One might have expected to find that, say, melody strongly favors a single central pitch, and meanders away from this pitch as if tied to it by an elastic band, in which case the expected distribution would be disproportionately found on or near that pitch, and would fall steeply lower and lower the farther away a pitch is from that central one. Melody does not, however, behave like this. Melody is more Doppler-like, sampling pitches within its home range in a fairly egalitarian fashion.

Both Doppler and melodic pitches have a fixed home range, and both tend to roam fairly uniformly over their home. Let's now ask whether some regions within the home range are "stickier." That is, are there regions of the tessitura where, if the melody goes there, it takes longer to get out? This differs from what we just finished discussing—that concerned the *total* amount of time (actually, the total number of notes) spent in regions of the tessitura, whereas we are now asking how long in duration each singular visit to a region tends to be. To understand sticky pitches in melody, we look to the Doppler shifts of movers and ask whether any Doppler pitches are sticky.

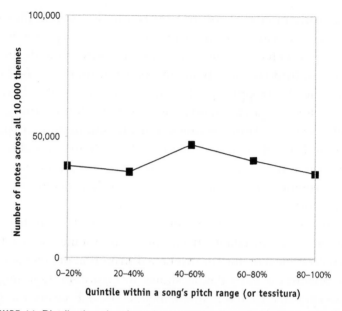

**FIGURE 44.** Distribution showing where within a theme's tessitura notes tend to lie, among the nearly 200,000 notes in the 10,000 classical themes. (That is, for each note, its position within its theme's tessitura was determined and its quintile recorded. The plot shows the distribution of these values.) One can see that themes tend to have pitches sampling widely across the tessitura, with little tendency to favor some parts of the tessitura over others. (The shape looks identical if the distribution for each theme is separately determined, and all 10,000 distributions averaged, and the error bars on such a plot are far too small to discern.) Note again that this analysis uses the tessitura for each piece, rather than measuring the number of semitones away from the average pitch in the song; the latter analysis would lead to a more normal distribution, falling quickly in probability away from the average, something researchers Tierney, Russo and Patel found in 2008. In light of the result here, the normal distribution they found is due to the distribution of tessitura widths, not the distribution of pitches within the tessitura.

There are, indeed, sticky Doppler pitches; they are the pitches near the top and bottom of the pitch range. To see why, imagine again a mover who is running in circles out in front of you, as depicted in Figure 45. Even though the mover is going through all directions uniformly, the pitches tend to change most quickly when the mover is whizzing horizontally by, either dropping quickly in pitch when passing nearby, or rising quickly in pitch when whizzing by at the far side of the circular path. When the mover is in the approach or withdrawal parts of the path, on the other hand, the pitch is fairly stable and high or low, respectively. Figure 45 shows these four segments of the circular path of the mover, and one can see that the pitch in the "toward" and "away" segments is much more stable than in the two "whiz by" segments. Doppler pitches vary less quickly near the top and bottom of their home range. The prediction, then, is that melodic pitch tends to change more slowly near the top and bottom of the tessitura. Does it? To test this, Sean Barnett measured the durations for all notes among the 10,000 classical themes. Each note was classified as a bottom, intermediate, or top note for its theme, and the average duration was computed for each of the three categories. Figure 46 shows these average durations, and one can see that lowest and highest notes in themes tend to be 17 percent longer in duration than notes with intermediate pitches.

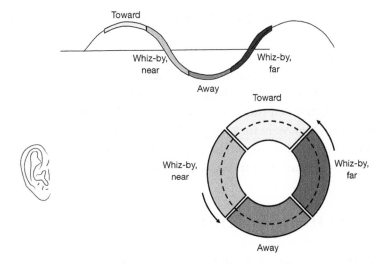

**FIGURE 45.** Doppler pitch changes slowly when near the maximum and nimum of the Doppler pitch range. Melodies also share this, as shown in Figure 7.

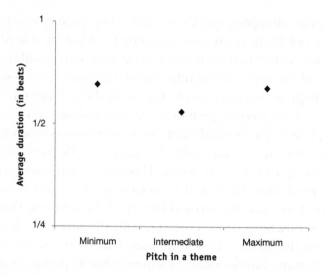

**FIGURE 46.** Across the 10,000 classical themes, this plot shows the average duration of the minimum, intermediate, and maximum pitch in a theme. One can see that the minimum and maximum pitch in a theme each tend to be longer in duration than intermediate pitches. (Averages of durations were computed in logarithmic space. Error bars are too small to see.)

In this section we looked at three facets of melody's home range. We saw that (i) melodies typically *have* a fixed home range, called the tessitura; (ii) melodies tend to distribute themselves fairly uniformly within their home range; and (iii) melodies tend to dwell longer at the edges of their range. Melody behaves in these ways, I am suggesting, because Doppler shifts behave in these ways. Melody is broadly Doppler-like in the home it keeps and the manner in which it distributes its movements and time throughout its home.

# 6 FAST TEMPO, WIDE PITCH

In the previous section we examined melody's home range— its size, and melody's hangouts within it. One facet of melody we discussed was that it tends to remain in a cage, called the tessitura, and I am suggesting that the top and bottom of the tessitura correspond to the Doppler pitches when the fictional mover is directed toward and away from you, respectively. But remember that Doppler shifts are greater when the mover has greater speed. A car driving past you at a crawl will have a small difference between its high approaching pitch and its low moving-away pitch. But if you stand at the side of the freeway the difference in pitch as the cars pass you will be much greater. It follows from these simple observations that faster-tempo pieces of music should have bigger home ranges for their melodies. That is, if melodic contour has been culturally selected to mimic the Doppler shifts of movers, then the prediction is that music with a faster tempo (more beats per minute) should have a wider tessitura.

To test this, Sean Barnett and I measured the tempo and tessitura width of the melodies of all the pieces in the *Classical Fake Book* (Hal Leonard Corp.). (We did not use the *Dictionary of Musical Themes* here because it does not include tempo data.) Figure 47 shows how tessitura width varies with tempo (for just those pieces originally intended for keyboard). One can see that although tessitura width does not change for the several low tempos, it rises among the faster tempos. Tessitura width increases with greater tempo, as predicted from the fact that the Doppler pitch home range widens as mover speed increases. This is particularly striking because themes with wider tessituras tend to be more difficult to play, and so one might predict that wider-tessitura music would go with a *slower* tempo, but this is the opposite of what we in fact find.

One might wonder whether this result could be due, instead, to a general phenomenon in which faster-tempo music tends simply to amplify musical qualities, whatever they may be. Caitlin Morris measured the range of loudness levels—the "loudness-tessitura" width—and the tempo for a sample of 55 pieces in Denes Agay's piano anthology, *The Classical Period*. Figure 48 shows how the width of the loudness range varies

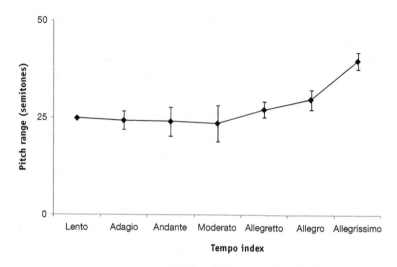

**FIGURE 47.** Tessitura (of melody) width versus tempo, among all 92 pieces for keyboard in the *Classical Fake Book* for which tempo data could be acquired. One can see that faster music tends to have wider tessituras, consistent with the Doppler interpretation of melodic pitch. (We found the same result when we used the data for all pieces.) The *Classical Fake Book* was used for two reasons. First, it is helpful because fake books are not cluttered with the notes from the chords (chords are notated via letter labels). Second, it is the only *classical fake book I* possess, so it amounted to an easy-to-get, unbiased sample.

with tempo, and one can see that there is no trend. The pitch tessitura width does not, then, increase in Figure 47 merely because of some general proclivity to amplify musical qualities at higher tempos. In fact, the *lack* of change in "loudness-tessitura" width as a function of tempo is something the music-is-movement theory *does* predict, assuming that loudness in music is primarily driven by proximity, as we discussed in detail in the "Nearness versus Stompiness" section of Chapter 4. Imagine that a mover carries out a bout of behavior in your vicinity at low speed. Now imagine this mover is asked to repeat the same bout of behavior, but this time moving much more quickly—that is, at a higher tempo. In each case the mover is, we presume, going through the same sequence of spatial coordinates, and thus the same sequence of distances from the listener. And so it immediately follows that the mover courses through the same sequence of loudnesses no matter whether moving slowly or quickly. The music-is-movement theory predicts, then, that unlike pitch, the range of loudnesses should *not* change as a function of the music's tempo—faster music, *same* loudness range—and that's what we found.

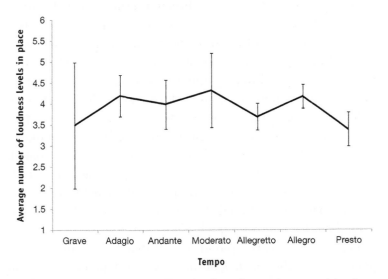

**FIGURE 48.** "Loudness-tessitura" width (i.e., the total range of loudness levels) versus tempo, sampled from 55 pieces in Denes Agay's *An Anthology of Piano Music, Vol. II: The Classical Period* (used instead of the Dictionary of Musical Themes because the latter does not possess loudness information). Unlike **(pitch)** tessitura width which is expected—and does—increase with increasing tempo, the loudness-tessitura is expected—and does—remain constant. This anthology was chosen because it was the only proper non-fake, non-lesson book I possessed at home.

We see, then that faster-tempo music behaves like faster-tempo movers: in each case the range of pitches increases with tempo, and the range of loudnesses does not change. Essentially, these results show us that the physics of movers is found in the structure of music. The upcoming Encore section continues the search for physics in music, and concerns momentum and Newton's First Law.

# 7 NEWTON'S FIRST LAW OF MUSIC

Objects at rest stay at rest unless pushed. And objects moving continue moving in the same direction unless pushed. This is Newton's First Law of Motion, which concerns inertia. This is a fundamental law of physics, and applies to any object with mass. Humans have mass, so it applies to us as well. And if music sounds like hulking humans moving about, then even music should adhere to Newton's First Law of Motion. Does it?

Before attempting to answer this, let's make sure we steer clear of one of the psychological handicaps I talked about in Chapter 4: the tendency to interpret musical pitch as spatial. As musical notes rise and fall on the page, or as your hands move hither and thither on the piano, it's hard to resist the feeling that inertia should show up in the musical domain as a tendency for a moving pitch to keep on moving in the same direction. But this is a pitfall. Recall that I am claiming that pitch is about the *direction* of the mover, not about position in space. Changes in pitch are therefore about changes in the mover's direction, not about changes of position in space.

With our memory jogged about the meaning of pitch, what are the expected musical consequences of Newton's First Law? A change in melody's pitch means the fictional mover's direction of motion has changed. Let's ask ourselves, then: if a mover changes direction, is there any physical tendency for the mover to *continue* changing direction? Purely physically, Newton's First Law tells us *no*. Any subsequent turn would require yet more force, without which the mover will continue going in whatever direction it *was* going. When a moving object for some reason makes a 30-degree change in direction,

the inertial tendency is precisely *not* to continue turning, but to continue going straight in the new direction. The same is true if a change in Doppler pitch is due to a change in speed— a change in speed does not lead to a further change in speed —but I'll always presume movers are staying at constant speed, the relevant case for music at constant tempo.)

The pitch of a mover, then, following the physics of movement itself, tends to stay the same. And if the pitch *does* change, it will have a tendency to stay at the *new* pitch – the mover's new direction —*not* to continue changing pitch. Newton's First Law for the pitches of movers is, therefore, that pitches have *no* inertia. Inertia is about how spatial changes tend to continue, *not* about how velocity (speed and direction) changes tend to continue. And because pitch is about velocity (i.e., speed and direction), not spatial location, pitch changes do not tend to continue. (If one *were* to imagine a spatial metaphor for how pitch changes, it would be movement of a bead in thick syrup: it moves if pushed with a fork, but *immediately* halts when no longer pushed.)

If melody's pitch contour acts like the Doppler pitches of a mover, then musical pitch is expected to have *no* "inertia" to continue moving *in the same direction*—"up" or "down". I had Sean Barnett carry out an analysis oon the entire data set of 10,000 classical themes, and we found that indeed, there was little or no inertia for pitch, just as is expected if melodic pitch contours sound like Doppler pitches from moving humans. In particular, for a one-semitone change, the probability of continuing up after a semitone up was 49.14 percent, and the probability of continuing down after a semitone down was 51.33 percent. For two semitones, the values were 47.06 percent and 56.17 percent. Pitches therefore do not act like spatial location: if pitch were spatial, then a change in pitch *would* tend to lead to more of the same kind of change due to inertia, and those percentages I just mentioned should have all been much greater than 50 percent. Instead, and as predicted,

pitches have little or no tendency to continue changing the same way they have been. Pitches act like Doppler shifts, following the expectations of Newton's First Law of Motion by *not* exhibiting pitch inertia (because inertia does not apply to directions of motion).

Although our data showed no strong bias towards pitch changes continuing in the same direction (which is the signature of true spatial momentum), note that there was, for both one- and two-semitone changes, a *slightly* greater tendency for pitches to go down—a small degree of downward momentum. To further examine this, I need to discuss some subtleties I have glossed over so far.

The ecological interpretation of pitch is the mover's direction of motion, but more carefully expressed, it is the mover's direction of motion *relative to the listener*. With this in mind, there are actually *two* fundamentally different ways for a mover's Doppler pitch to change. The first is what I have assumed in this section thus far: the mover turns. But Doppler pitch *can* change even when the mover does not turn, and this second source of changing Doppler pitch you are very familiar with, because it happens every time a mover passes you, including the generic passing train. When movers pass listeners, their pitch falls. In fact, whenever an object simply moves *straight* its pitch falls (unless the object is directed *perfectly* toward or away from the listener). It is not, then, quite right to say that Doppler pitches have no pitch momentum. Straight-moving movers have falling pitch, and straight-moving movers tend to keep going straight (because of inertia), and therefore falling pitch in *these* circumstances *does* tend to keep on falling.

If only we could tell the difference between the pitch changes due to a mover actually turning and the pitch changes due to a mover simply going straight! We could then predict a lack of pitch momentum for the former, but predict the presence of pitch momentum for the latter. Actually, we *can* tell them apart. When a mover turns, it is intentional and occurs fairly

quickly (on average about 45 degrees per step, as discussed earlier in the section of Chapter 4 called "Human Curves"), *and* it can be a change in direction either *more* or *less* toward the listener. Intentional turning behavior therefore tends to lead to large pitch changes that can be upward or downward. But the second source of Doppler pitch change is the one due to movers going straight (and going by). In this case there tend to be a *lot* of steps over which pitch falls—because now the falling pitch is, in essence, due to continuous change of position in space—and so the pitch change per step is small, *and* is always downward.

Here, then, is how we can distinguish the straight-moving mover from the turning mover. When pitch falls by only a small amount, it tends to be the signature of a straight-moving mover passing. But when the pitch change is *not* consistently small and downward, it is typically due to the mover turning. Thus, a turning mover is given away by either of two pitch cues: (i) a large pitch change, whether upward or downward, implicates a turning mover, and (ii) any pitch change upward at all, small or large, implicates a turning mover (because straight-moving movers only have falling pitch, not rising pitch).

We do therefore expect Doppler pitch to possess inertia in just one circumstance: when pitch falls by a small amount. Small drops in pitch are more probably attributable to a straight-moving mover. Because straight movers have inertia and are thus likely to continue moving straight, small pitch drops *do* tend to have inertia. Small pitch upswings do not, however, have inertia, and neither do large pitch changes, whether up or down.

Is this what we find in music?

We already saw evidence for this earlier in this section. Recall that there was generally little or no inertia for pitch—the probability of a pitch change continuing in the same direction was near 50 percent. But let's look at the pitch momentum numbers again, more carefully this time. For a semitone pitch

change, the probability of continuing in the same direction was 49.14 percent and 51.33 percent for upward and downward, respectively. (Their standard errors are small—0.005 and 0.004, respectively—because these are averages across many thousands of instances.) The same asymmetry was found when considering whole-step changes of pitch, but now with respective values of 47.06 percent and 56.17 percent. (Standard errors are each 0.004.) These results are consistent with those of Paul von Hippel that David Huron discusses in his book *Sweet Anticipation*: no momentum following small steps upwards, but significant momentum downwards. The signature of pitch momentum is a value greater than 50 percent, and only the downward pitch change has this. (Upward pitch change is below 50 percent, meaning that a little more than half of the time a semitone upward is followed by either no change in pitch or a downward change in pitch). For larger pitch changes, we found that neither upward nor downward pitch changes had any pitch momentum (i.e., the probability of continuing to change in the same direction was below 50 percent). Not only, then, does melodic pitch contour have a counterintuitive tendency to have no inertia, like the pitches of movers—but it breaks this tendency exactly when movers do. Consistent with melody's meaning coming from the Doppler shifts of movers, melody conforms to Newton's law of inertia.

In addition to the issue of whether pitch changes continue to change in the same direction, we can make a simpler observation. Let's ask ourselves what the baseline expectation is for the Doppler pitch change of a mover. One's first intuition might be that, in the absence of any information otherwise, we should expect a mover's Doppler pitch to remain unchanged from one step to the next. Doppler pitches, however, do not typically hold still. Instead, the most fundamental baseline expectation (inertia) is that movers continue moving in whatever direction they *were* going. People tend (though not as strongly as inanimate objects!) to keep going straight, and

thus the baseline, or generic expectation, for pitch change is that pitch *falls*, and generally by a small amount (compared to the pitch changes of a turner).

Now consider another observation about movers. Suppose that a mover is carrying out bouts of behavior around you, *and* is directing those behaviors toward you. Notice that this mover will have to make intentional turns toward you to keep orienting his behavior toward you. But also notice that he or she never has to deliberately turn *away* from you. This is because once a mover is directed roughly but not exactly toward you, going straight inevitably leads to a movement like veering away. Simply going straight will cause the mover to pass by you and depart. Turning is necessary to go toward someone, but not to go away. (This is essentially because a listener is in one location, and all the other locations are where that listener is *not*.) Thus, if musical melody is about listener-directed bouts of behavior, not only do we expect small pitch changes to more commonly be downward, we expect large pitch changes to more commonly be upward.

Do we find this in music? Do we find that melodic contours have a general tendency to fall gradually? And do we find that pitch drops tend to be smaller than pitch rises? Piet G. Vos and Jim M. Troost of the University of Nijmegen indeed found this among a sample from the *Dictionary of Musical Themes*. We carried out our own measurements over the entire data set: Sean Barnett measured the relative probability that a pitch changes upward (so that a probability greater than 0.5 means an upward tendency, and a probability less than 0.5 means a downward tendency) as a function of the size of the pitch change. One can see these results in Figure 49. For small pitch changes, the probabilities are mostly below 0.5, meaning that small pitch changes tend to be downward, as expected. And for large pitch changes the probabilities are mostly above 0.5, meaning that large pitch changes tend to be upward, also as expected.

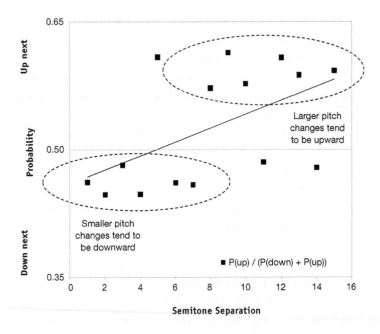

**FIGURE 49.** The y-axis shows the relative probability that a pitch will go up on the next note, among the 10,000 classical themes in Barlow and Morgenstern's Dictionary. A value of 0.5 means it is equally likely to go up or down in pitch. The x-axis represents how far the pitch changes (in number of semitones). One can see that for small pitch changes the probability tends to be below 0.5, meaning pitches tend to fall. But for large pitch changes, they tend to rise. Sean Barnett, then a graduate student at RPI, made these measurements.

Another characteristic difference between rising and falling Doppler pitches from movers is that when a mover passes by, going straight, the pitch doesn't just fall slowly with "inertia," but *continues* to fall over an *extended portion* of the pitch range. The train that has reached your position has, for example, dropped from its maximum pitch to its baseline pitch, and will then drop through the lower half of the pitch range as it goes past and away. These kinds of long Doppler pitch runs, then, are more commonly downward than upward for movers in the physical world. Are long pitch runs for *melodies* more commonly downward than upward? Sean Barnett

measured runs among the 10,000 classical themes. In particular, he recorded runs spanning the bottom or top half of the
theme's tessitura. Setting a low bar for what counts as a run—
two or more notes approximately filling (more than 80 percent) the upper or lower half of the tessitura—51.86 percent
of the 212,542 runs were downward. A two-note run is not
very run-like, and our expectation is that if we create a more
stringent standard for what counts as a run, then we should
find an even greater asymmetry between up and down, with
an even greater share of runs being down. Indeed, when Sean
required a run to have five or more notes in the same direction, 54.22 percent of the 11,119 runs that qualified were in
the downward direction.

Consider now yet another ecological regularity in this vein.
Let's ask ourselves: Are these falling-pitch runs due to straight-
moving movers more likely to occur when movers are near or
far? When a mover is far away, in order for that mover to implement a long downward run, the mover must continue straight
for a great many steps without turning. It is quite likely that
the mover will turn somewhere over the course of that long
walk. But if the mover is close by, the mover need only move
straight for a relatively small number of steps to engender a
substantial downward pitch run. Big downward Doppler pitch
runs are therefore disproportionately probable when near. Do
we find this in music? As we discussed in Chapter 4, distance
from the listener is encoded in music by loudness, and so our
expectation here is that louder segments of music (i.e., passages depicting a more proximal mover) are more likely to
have good-sized downward pitch runs. RPI graduate student
Romann Weber measured runs spanning at least half the tessitura from Denes Agay's *The Classical Period,* and calculated
the probability that such pitch runs are downward as a function of loudness. As can be seen in Figure 50, the probability
of a large downward pitch run rises with loudness, consistent
with the ecological expectation.

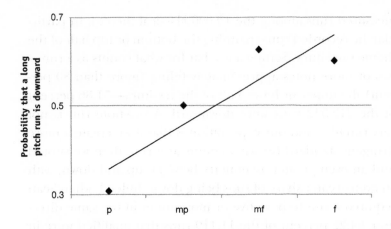

**FIGURE 50.** Pitch runs spanning at least half the tessitura width among a sample of 37 pieces from Denes Agay's *An Anthology of Piano Music, Vol. II: The Classical Period.* Forty such runs were found, the loudness during the run measured, and the relative probability that the run is up or down computed. Louder segments of music have a greater probability of long downward runs, consistent with expectations from human movement.

Newton's First Law is found in music where it should be found if melodic pitch is about Doppler shifts. Melodic pitch acts like a mover's direction, and thus has *no* pitch momentum, just as inertia predicts. But as we looked more closely, we realized that there are fundamental asymmetries between upward and downward Doppler pitch changes, asymmetries also found in melody. Melody does show pitch momentum in the special case of small downward changes in pitch, as expected from the dynamics of movers. And melody generally drifts downward gradually, as expected from the fact that all straight-moving movers have slowly falling pitch (unless moving directly toward or away from the listener). And melody takes larger jumps upward than downward in pitch, also something expected from movers orienting their bouts of behavior toward you. Melody also favors longer runs downward than upward, something we also expected from the sounds of movers. And finally, like closer movers, louder segments of music tend to have disproportionately more large downward pitch runs.

The Encore sections thus far have mostly concerned rhythm and melody. Loudness did come up in Encores 4, 6, and 7. The next and last Encore section is about loudness, providing further evidence that loudness in music behaves like loudness due to the proximity of the mover.

# 8 MEDIUM ENCOUNTERS

In the Chapter 4 section titled "Slow Loudness, Fast Pitch," we saw that loudness varies slowly, consistent with the time scales required for movers to vary their distance from you, the listener. We must be more careful, though. If a mover were a "close talker," tending to move about uncomfortably close to you, then even small changes in distance could lead to large changes in loudness, due to the inverse square law for loudness and distance. But in real life, more than close encounters, we tend to have medium encounters: the movers we typically listen to tend to be in the several- to ten-meter range, not in the centimeter range, and not in the tens or hundreds of meters range. At "medium" distances, large loudness modulations don't occur over just one or several steps. They require more steps, plausibly in the range of the approximately 10 beats we found for the average loudness duration in Chapter 4.

Not only are our experiences of movers usually at a "medium" distance, but it seems reasonable to expect that individual bouts of behavior tend to occur at an average "medium" distance. Recall our generic encounters from the section titled "Musical Encounters" in Chapter 4: the "center of mass" of the A-B-C-D cycle of movement would be representative of the average distance of a generic encounter. We see, then, that loudnesses of movers will tend to have a *typical* value. We therefore expect any piece of music to have a baseline loudness level it spends a disproportionate amount of time at, spending less time at loudness levels farther away from this average. Unlike Doppler pitches, which have a distribution that is fairly broad and flat, the distribution of mover loudnesses tends to be more peaked. Is music like this? Does music spend most of its time at an average loudness level, relatively rarely venture out of that loudness zone, and more rarely still pursue greater loudness deviations from the average? Music is *indeed* roughly like this. Music tends to use *mezzo forte* as this baseline, with lesser and greater loudness levels happening progressively more rarely. RPI students Caitlin Morris and Eric Jordan measured the average percentage of a song spent at each of its loudness levels, and the results are shown in Figure 51. One can see that there is a strong "mountain" shape to the plot: pieces tend to spend more time at intermediate loudness levels than at loudness levels deviating far from the central values. (Although our data were broadly consistent with our expectation, there was a slight downward divot at *mezzo forte* relative to *piano* and *forte*, with the greatest percentage of time spent in *piano*.)

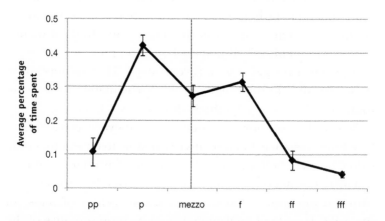

**FIGURE 51.** For each song, the total percentage of time spent at each loudness level was determined. These distributions were then averaged together across 43 pieces in Denes Agay's *An Anthology of Piano Music, Vol. II: The Classical Period.*

We can say more. Consider the obvious fact that there is less real estate—less space—near you than far from you. This asymmetry means that a mover has more chances to be farther than average from you than to be nearer than average to you. There should not only be, then, a roughly mountain shape to Figure 13, but the below-average levels of loudness should be more common than the above-average levels of loudness. The mountain should have a higher peak at lower-than-average levels of loudness. The distribution we just plotted in Figure 51 *has*, in fact, this expected asymmetry.

We can say something further still. Not only should movers spend a greater proportion of their time relatively far away than relatively nearby, but when they *do* get near, and thus relatively loud, this should be more transient. Why? Because the mover will more quickly leave the near region, for the simple reason that "the near" is an inherently smaller piece of land than "the far." This is indeed the case, as shown in Figure 52, also obtained by Caitlin Morris and Eric Jordan

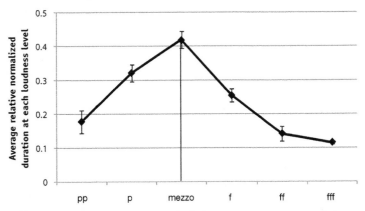

**FIGURE 52.** For each song, the average duration of each loudness level was computed, and then these per-song average normalized so that the sum across the levels equaled one. Then, these were averaged across 43 pieces measured in Denes Agay's *An Anthology of Piano Music, Vol. II: The Classical Period* One can see the asymmetry. As predicted from the spatial asymmetries of near and far, music should tend to have longer durations at lower-than-average loudness levels compared to higherthan- average loudness levels.

We see, then, that loudnesses distribute themselves as expected if they are about proximity. Encounters have a typical distance; more cumulative time is spent farther than nearer; and nearer segments of encounters tend to be short-lived relative to farther segments.